T0245399

Modernising solid waste management at municipal level

Modernising solid waste management at municipal level

Institutional arrangements in urban centres of East Africa

Christine Majale Liyala

Environmental Policy Series – Volume 3

Wageningen Academic
P u b l i s h e r s

ISBN: 978-90-8686-189-7
e-ISBN: 978-90-8686-745-5
DOI: 10.3920/978-90-8686-745-5

First published, 2011

© Wageningen Academic Publishers
The Netherlands, 2011

Wageningen Academic Publishers
P.O. Box 220
6700 AE Wageningen
The Netherlands
www.WageningenAcademic.com
copyright@WageningenAcademic.com

Preface

The task of municipal problem solving has become a team sport that has spilled beyond the borders of government agencies and now engages a far more extensive network of social actors - public as well as private, non-profit and profit. Solid waste management is one of the key tasks associated with municipal authorities. It is of particular interest because of its flexibility compared to other services at the municipal level. This means the number of options for addressing solid waste management at municipal level are considered many.

The Lake Victoria Basin allows a comparison of institutions and practices in the different urban centres that are found in the basin. The ultimate aim has been to advance plausible options for institutional arrangement to improve solid waste management at the benefit of the urban poor in the Lake Basin in East Africa.

This dissertation is the result of five years of research work done under the PROVIDE project (working on sustainable urban infrastructures in cities of the Lake Victoria Basin, East Africa) with funding from INREF.

I extend my sincere gratitude to my promoter Professor Gert Spaargaren and my co promoter Dr. Peter Oosterveer who have patiently guided me during the research process, providing me with feedback and comments. I thank Dr. Caleb Mireri who also took part in assisting me to shape the research.

My gratitude goes to the other staff of the Environmental Policy Group for their assistance and guidance at various stages and levels. Prof. Arthur Mol, Dr. Bas van Vliet and Dr. Szanto Gabor for all the logistical support and guidance, Loes Maas from the Methodology group for her tireless assistance in shaping the methodology of the study, thank you!,

Thanks to Corry Rothuizen for her administrative and logistic arrangements and support during the whole time I was at Wageningen University and even when back home in Kenya. Thank you!

I also thank the PROVIDE group: Aisa Solomon, Judith Tukahirwa, Sammy Letema, Fred Owegi, Meshack Katusiimeh, Richard Oyoo, Fredrick Salukele and Tobias Bigambo. You are an amazing team. Not forgetting the following: Dorien Korbee, Hilde Toonen, Judith van Leeuwen Elizabeth Sargant, Leah Ombis, Harry Dabban, Eira Carballos, Carolina Marciel, Sarah Stattman Kanang Kantamaturapoj, Judith Floor, Vijge Marjanneke and Jennifer Lenhart for their company and assistance. I thank Dorien Korbee and Hilde Toonen for translating the English summary into Dutch in a short notice.

Next, I must thank my parents, Wycliffe and Eunice Majale. You have provided encouragement in my academic pursuits. You have also helped make this possible by being there to do whatever needed to be done, especially taking over the parental duties of looking after Benaya. Mum Philgonna Mulamba, you have also have also been a source of support and help for which I am grateful. Thank you for being there and for the encouragement you have provided along the way. To the rest of the family, thank you!

I thank my dear husband and best friend, Reuben. There are not adequate words to describe the support you have provided. There would have been no way this accomplishment would have been possible without your unwavering encouragement, love and commitment. We share this degree as we share everything else in the life we have built together. This life now includes our son, Benaya, who provided a new motivation to finish this dissertation. Benaya, you are a source of inspiration and awe in my life and this degree is shared with you as well.

I would also like to thank my contact persons and interview participants for the access each of you granted me into your experiences in Kisumu, Jinja, Mwanza, Kisii, Homabay and Migori. Hopefully, it will start a dialogue and spark future studies that will inform the theoretical debate surrounding governance in municipal service provision in East Africa.

Last and most importantly, I must thank my Lord and Savior, Jesus Christ. For, I can do all things through Christ who strengthens me - Philippians 4:13.

Table of contents

Abbreviations

CBD	Central business district
CBO	Community based organisation
COMESA	Common market for Eastern and Southern Africa
CSO	Civil society organisation
EAC	East African community
EMT	Ecological modernisation theory
FBO	Faith based organisation
FOCJ	Functional, overlapping, competing jurisdictions
GPT	Graduated personal tax
IGAD	Intergovernmental authority on development
ILO	International labour organisation
IMC	Inter municipal cooperation
KIWAMA	Kisumu waste managers association
Kshs	Kenya shillings
LASDAP	Local authority service delivery plan
LATF	Local authority transfer fund
LC	Local council
LG	Local government
LGA	Local government authority
LVRLAC	Lake Victoria region local authorities cooperation
LVB	Lake Victoria basin
LVBC	Lake Victoria basin commission
LVWATSAN	Lake Victoria water and sanitation program
MASMA	Mwanza solid waste management association
MLG	Multi-level governance
MMA	Modernised mixtures approach
MOU	Memorando of understanding
MSF	Multi-stakeholder forum
NEMA	National environment management authority
NEMC	National environment management council
NFP	National focal point
NGO	Non-governmental organisation
OECD	Organisation for economic cooperation and development
PA	Provincial administration
PHI	Public health inspector
PPP	Public private partnership
PROVIDE	Partnership for research on viable infrastructure development
RCC	Refuse collection charge
ROC	Regional organisation of councils
SACCO	Savings and credit cooperative

SAP	Structural adjustment program
SIDA	Swedish international development agency
SWM	Solid waste management
Tshs	Tanzania shillings
Ugshs	Uganda shillings
UNIDO	United Nations development industrial organisation
UNEP	United Nations environmental programme
USD	United States dollar

Chapter 1.
Introduction

1.1 Urban authorities and solid waste management

Waste as a subject is certainly part of a growing discourse attracting the attention of anthropologists, economists, historians, sociologists, amongst others, most of whom point out that waste is an indication of the negative side of different dichotomies such as efficiency/inefficiency; usefulness/uselessness; order/disorder; gain/loss; clean/dirty; alive/dead; fertile/sterile. Sociologist Gille in her book - *From the cult of waste to the trash heap of history* - speaks of waste proving to be a good lens through which social scientists can get a glimpse of other underlying social and cultural anxieties and moral dilemmas. Anthropologist Douglas in her book - *Purity and danger* - speaks of dirt as an indication of disorder, that is 'matter out place'. In the same tone, scholars who have written on governance speak of waste management as being one of the most visible urban services whose effective and sustainable management serves as an indicator for good local governance, sound municipal management and successful urban reforms (UN-Habitat 2010; Van Dijk, 2008; Van Dijk and Oduro-Kwarteng, 2007). Waste management is thus very much connected to the performance of municipalities.

Urban authorities in most developing countries are the institutions generally responsible for the provision of solid waste collection and disposal services. A solid waste crisis therefore can significantly undermine the credibility of an urban authority. According to UN-Habitat (2010), solid waste management may not be the biggest vote-winner, but it has the capacity to become a full-scale crisis, and a definite vote-loser, if things go wrong.

Yet for the urban centres in East Africa, the storage, collection, transporting, treatment, and disposal of solid wastes, particularly wastes generated in medium and large urban centres, are reported to have become a relatively difficult problem. In most urban areas, only a fraction of the waste generated daily is collected and safely disposed off. In Nairobi for instance, generation rates as of 2009 are estimated at 1,850 tons/day of which just about 33%[1] is collected and gets to the dumpsites (JICA, 2010). In Kampala, of the 1,300 tons generated daily, just about 21.7% is collected (Okot-Okum, 2006). Where people have stored waste prior to collection using non-biodegradable material instead of the skips provided by councils, this material ends up at the dumpsite with the waste.[2] Collection of solid waste is usually confined to the city centre and high income neighbourhoods, and even there the service is usually irregular. The urban poor - often residing in informal settlements with little or no access to solid waste collection and often in areas that are contiguous with open dumps - are particularly vulnerable. Most parts of cities therefore never benefit from public waste disposal. Consequently, most urban residents and operators resort to burying or burning their waste or disposing of it haphazardly. The capacity of urban authorities to process, or re-use solid waste in a cost-efficient and safe manner is reported to be

---

[1] UN-Habitat (2010) reports the collection estimates to be 60 to 70%, but most of this is by the private collectors and does not get to the dumpsite.

[2] See Okot-Okum (2006) for describing the case of Kampala.

far more limited (Karanja, 2005). This situation brings to the fore the question of institutional arrangement for solid waste management (SWM) at the municipal level.

A number of academic studies have been done on solid waste management in Africa including: Awortwi (2003) who focuses on governance in multiple arrangements and the relationship between capacity and contractual arrangements. Obirih-Opake (2002) focuses on the impact of decentralisation and private sector participation on urban environmental management. Karanja (2005) focuses on solid waste management and sustainable development issues, identifies different actors and institutional arrangements, looking at the role and interests of different actors, their successes and failures.

This thesis will build on some of the arguments of these authors. It however, deviates slightly from these studies by adopting a bioregional perspective that covers the Lake Victoria Basin and using the Ecological Modernisation theory and the Modernised Mixtures approach. Thereby this thesis seeks to present feasible options for institutional arrangement for solid waste management at the municipal level.

The remainder of this chapter therefore presents a brief on the trends in SWM infrastructure provision; the theoretical basis that guides the study; the study context and the further outline of the thesis.

1.2 Trends in SWM infrastructure provision in East Africa

For the urban centres of East Africa, SWM under the local authorities has for a long time been centralised through the use of large scale infrastructure in service provision. The use of highly mechanised refuse collection trucks, which in most cases have been imported from industrialised countries (Karanja, 2005; Rotich *et al.*, 2006), has been common practice. These trucks cannot access most of the low income areas and experience frequent breakdowns due to lack of maintenance and repair (ADB, 2002). In other towns skips for waste collection can be found centrally located denying areas located far from the central business district access to this collection infrastructure. Infrastructure for waste disposal in most urban centres have been the centralised open dumps designated more for convenience of available space rather than because of accessibility and ecological sustainability considerations. Illegal dumpsites have therefore sprung up time and again. In essence, the large scale, centralised infrastructure provision has reinforced inequity in the distribution of SWM services other than perhaps the intended need of making it accessible to most people. The inadequacies of these large scale systems of SWM infrastructure provision by the local governments necessitated the move towards small scale infrastructure. Dating back to the 1970s, the use of appropriate technology was encouraged but this was more oriented towards waste re-use and recycling. With the growth in the roles of civil society and private firms in the 1980s, such small scale, flexible, low technology and decentralised approaches flourished more. Today, it is commonplace to read about small scale waste collection using wheelbarrows and pickups by CBOs and private firms, neighbourhood transfer stations and recycling and treatment practices like household or community waste composting initiatives. These small scale technologies however, offer only solutions to individual households and/or locations, and are developed where the finances, technological capabilities and organisational capacities are severely limited. According to Spaargaren *et al.* (2006), both users and local authorities consider such technologies 'low quality'.

Local authorities are thus faced with a dilemma of which path to choose for improving SWM as both the centralised and the decentralised systems show serious weaknesses. Attention should therefore be given towards exploring alternative modes of modernising the SWM system. Apart from the challenge of increasing the coverage of waste collection (to parts where there is less infrastructure and ability to pay is low), attention should also shift from merely removing waste before it becomes a health hazard to creatively minimising its environmental impact (Un-Habitat, 2010).

1.3 Ecological modernisation theory and modernised mixtures approach

Ecological Modernisation Theory (EMT) focuses on environmental reforms in social practices, institutional designs, and societal and policy discourses to safeguard societies' sustenance bases (Mol and Sonnenfeld, 2000). In SWM, EMT argues for improved performance in both economic and environmental dimensions (Scheinberg and Mol, 2010). It includes criteria for assessing ecological performance like the ones put forward by SWM research in OECD and transition countries: minimising the formation of contaminated water ('leachate') from 'sanitary landfills' to prevent its release into groundwater and surface water; working as much as possible with separation at source and recycling/re-use; closing nutrient cycles by capturing organic waste (see UN-Habitat, 2010). EMT also argues for the use of 'modern' institutions when working towards solutions where the market, technology and professional NGOs have key roles to play next to governments at their different levels/scales. Yet since EMT was developed in the context of OECD countries, it is not directly applicable to the East African context and this is where the Modernised Mixtures Approach (MMA) comes in.

Modernised Mixtures refer to sociotechnical configurations of infrastructures – in this case SWM infrastructures and services - in which a variety of features of (modernising) systems are deliberately and reflexively being constructed in response to the challenges created by a changing social, economic, and environmental context (Hegger, 2007; Oosterveer and Spaargaren, 2010; Scheinberg and Mol, 2010; Spaargaren *et al.*, 2006). Under the MMA, we have intelligent combinations (or mixtures) of simple and advanced technologies; small and large-scale systems; centralised and decentralised forms of control; public, private, formal and informal actors; citizen participation and professional management; and uniformity and diversity of systems. In essence, when working with modernised mixtures, one leaves behind the dichotomy dividing centralised, large-scale, high-tech solutions on the one hand from the decentralised, appropriate, small scale and low-technology solutions on the other (Spaargaren *et al.*, 2006). Instead of opposing centralised and decentralised paradigms, the best of both paradigms are to be combined into new configurations. Specific criteria under the MMA guide the assessment of different options (or combinations) optimised throughout waste flows, institutions and economics. They are:

- *Ecological sustainability* of the infrastructures and practices involved: to what extent do the new systems or the new technological options that become part of existing systems, improve the environmental performance of the urban infrastructure?
- *Institutional sustainability* concerns the extent to which a new system becomes embedded in existing socio-political and cultural systems at the local and national level, while improving their performance.

- *Accessibility* (particularly of the poor to avoid exclusion of particular groups): to what extent are specific groups included or excluded from environmental infrastructure due to financial, physical or cultural reasons?
- *Flexibility and resilience* (in both technological and institutional respect): how does the system or unit fit into more embracing future systems and how does it behave under different forms of instability (climatic, political, economic, institutional)?

This thesis focuses on institutional arrangements and uses the above criteria of ecological and institutional sustainability, accessibility and flexibility to present feasible options for SWM at municipal level. These criteria are operationalised in line with the different variables that are used in the study.

1.4 The question of institutional arrangements for SWM

How are SWM services actually delivered to people in developing countries today? The types of arrangement for service provision today range from self-provision through collective action independent of external agencies to indirect state provision through sub-contracting to other agencies – NGOs, private for profit companies, user groups among others. Generally, there is much agreement that monopolistic provision realised entirely through state agencies is unfeasible, undesirable, or simply rather old fashioned. However, there is little consensus on the alternative. Joshi and Moore (2004) argue that there is need to look beyond new discourses like New Public management and Public Private Partnerships indicating that the trend now is towards pragmatism, pluralism and adaptation to specific circumstances because the reality in developing countries is highly diverse. Some services, it is argued, cannot be effectively delivered to the ultimate recipients by state agencies because the environment is too complex or variable, and the costs of interacting with very large numbers of poor households are too high. In such cases, users become involved in an organised way at the local level. There are arrangements therefore that do not fit into standard categories. Some of these unorthodox arrangements are of recent origin, and are seen to constitute (smart) adaptations to prevailing local circumstances. They are widespread in developing countries but they raise many issues. Being cognizant of these diverse issues that come into play between what is considered standard and that which does not fit into such standard category, perhaps the next question would be what the most feasible institutional arrangements for SWM are at the municipal level? To help answer this, the following specific research questions are defined to guide the study:

1. What is the current status of the (physical) environmental infrastructures and the level of service provision for SWM in three designated urban centres in EA namely Kisumu (Kenya); Mwanza (Tanzania) and Jinja (Uganda)?
2. What are the existing policy arrangement for SWM in these three urban centres and what can be learnt from the differences and/or similarities amongst them?
3. What are the possibilities for cooperation in solid waste management amongst small neighbouring municipalities in Kenya namely Kisii, Homabay and Migori municipalities?
4. What is the role of regional organisations and networks in enhancing cross border infrastructure provision in solid waste management amongst municipalities in the lake basin?

1.5 Study context

The study is part of a bigger project – Partnership For Research On Viable Infrastructure Development (PROVIDE) – in East Africa that covers the Lake Victoria Basin and seeks to improve sanitation and solid waste management in the basin as part of achieving MDG7. MDG 7 intends to ensure environmental sustainability and two of its targets are particularly relevant for the PROVIDE project:

- *Target 10:* have by 2015, the proportion of people without sustainable access to safe drinking water and basic sanitation;
- *Target 11:* have achieved by 2020 a significant improvement in the lives of at least 100 million slum dwellers.

Key requirements to address the MDGs in general are: improved governance, technological innovation and diffusion, and enhanced financing mechanisms. Achieving targets 10 and 11, therefore also needs an integrated approach and to include contributions from political and social sciences, technology and economics. Different studies in the PROVIDE project are covering different aspects (technical, social, economic and governance issues) of sanitation and solid waste at different levels (household, neighbourhood, municipal, national and regional scales). The studies are also covering different urban centres in the East African Region. This study as earlier mentioned deals with solid waste management, emphasising governance issues at the municipal level. Jinja, Mwanza and Kisumu are the main urban centres under study, but the second part of the study also covers small municipal authorities (Kisii, Homabay and Migori) on the Kenyan side of Lake Victoria.

1.5.1 Profiles of the main urban centres

The three urban authorities that form the core of the study area are Kisumu (Kenya), Jinja (Uganda) and Mwanza (Tanzania). This section gives a brief description of their geographical location within the Lake Victoria Basin; their sizes and population figures.

Kisumu is the third largest urban centre in Kenya after Nairobi and Mombasa. It is located in Nyanza province in the Western part of Kenya. Geologically it sits on the arm of tertiary lava, which extends southwards overlooking the plains to the East and Winam gulf of Lake Victoria to the West. The lava formation is attributed to the tectonomagnetic activities associated with the Kano-Rift valley system. As a result the city is curved into a trough with the walls of the Nandi escarpment to the East dropping onto the floor of the Kano flood plains and gently flowing to the Dunga wetlands at the shores of the Lake Victoria. Kisumu covers an area of 297 km^2 of land mass and 120 km^2 under the lake. The population of the council has been increasing rapidly, and at a growth rate of 2.8% per annum it was estimated at about 500,000 in 2007 from 322,734 people in 1999.

Jinja is the second largest urban centre in Uganda after Kampala city. It is located 81 km East of Kampala. It is situated just north of the equator, on the northern shores of Lake Victoria and at the source of the Nile River. The town lies on a tapering plateau with an average altitude of 1230 meters above sea level. The municipality has an extensive shoreline in the east, south and west of both Lake Victoria and the voluminous waters of the Victoria Nile. It occupies an area of 28 km^2. It has a resident population of about 86,512 people (population census 2002) with a day population that doubles that figure due to peri-urban migrant labour. At a growth rate of 2.4% per annum, the population as of 2007 is estimated at 95,121 people.

Mwanza is the second largest urban centre in Tanzania after the city of Dar-es Salaam. It covers an area of 1,325 km^2 of which 425 km^2 is dry land and 900 km^2 is covered by water. Of the 425 km^2 dry land area, approximately 86.8 km^2 is urbanised while the remaining area consists of forested land, valleys, cultivated plains, grassy and undulating rocky hill areas. According to the 2002 National Census, Mwanza City had 476,646 people. With an annual natural growth rate of 3.2%, the population as of 2007, is estimated at 714,060 people.

The three were chosen because of certain similarities (and differences), but more so because they are all found on the shores of the lake basin as described above and they are all primary urban centres which makes them comparable in urban status.

1.6 Outline of thesis

The next chapter of this thesis, Chapter two, brings out a historical overview as well as the theoretical perspectives of the study. The historical part brings out the dimensions of change in the political landscape with reference to the place of local authorities, privatising municipal services, the role of civil society and regional integration. This allows the study to make arguments for institutional arrangements drawing from occurrences in the history and the changes that have taken place over time. The part that discusses the theoretical perspective seeks to show the effects of different institutional arrangements on decentralised public service provision and in turn on SWM. Therefore theoretical debates on centralisation vs. decentralisation and developmental state vs. network governance are presented. Multi-level governance is also discussed. These theoretical perspectives culminate into the conceptual frameworks that guides the empirical part of the study.

Chapter three is the first empirical chapter and it discusses the internal organisation within municipalities. It compares the performance of SWM tasks in technical and social respect amongst three municipalities-Jinja, Mwanza and Kisumu. In this chapter, the study also dedicates a brief section to explore municipal autonomy in SWM.

Chapter four compares the collaboration (formal and informal) between the municipal authorities and non-state actors in their respective jurisdictions. Arguments are made alongside the developmental vs. network governance debate.

Chapter five takes the study a step further by assessing opportunities for inter-municipal cooperation amongst three small neighbouring municipalities (Kisii, Homabay and Migori) on the Kenyan side of the Lake Victoria Basin. Discussions are made with reference to the multi-level governance discourse.

Chapter six presents the role of regional organisations and networks in cross country SWM at the municipal level. It narrows down to a study of two regional organisations, Lake Victoria Basin Commission and Lake Victoria Region Local Authorities Cooperation, which are categorised as statutory and voluntary regional arrangements respectively. The two organisations are assessed for their role in enhancing cross country cooperation in SWM.

Chapter seven is the concluding chapter and in answering the research questions formulated at the beginning of the study, it presents the conclusion of each part of the study. Towards the end, it presents the final observations.

Chapter 2.
Changing dynamics of politics in East Africa

2.1 Introduction

The existence and performance of municipalities in East Africa in environmental infrastructure and service provision of which solid waste management is one, has a historical component depicting changes that have occurred dating back to the 1960's when the three East African countries attained independence. Olowu (2002) explains that when these countries attained political independence with formal structures of democratic, representative government, political leaders in their bid to consolidate political power then opted for highly centralised modes of governance. This centralised mode of governance was reinforced by a culture of politics of patrimony in which all powers and resources flow from one source of power ('the father of the nation') to clients to shore up the regime. This pattern of power and resource distribution was strongly supported by both domestic and external actors until the late 1980s. The reasons adduced for adopting this approach included – rapid economic and social development actualised through centralised planning, unity and national integration, containment of corruption and political stability. In fact the argument was that if decentralisation would be necessary at all it must be in the form of administrative decentralisation or deconcentration - the sharing of responsibilities between central and local administrations which do not exercise any discretionary authority nor dispose of resources. Yet the 1990s marked an era of political and democratic approaches wherein decentralisation was progressively being seen (by governments, external actors and the increasingly influential civil society lobbies) as a means of enhancing democracy and citizen participation and (by governments and external actors) as a way of reducing the role, and in particular the expenditures, of the central government (Conyers, 2007). Over time, these changes have necessitated governmental reconfigurations, many of which have a powerful 'local' governance orientation. They include resurgent regional organisations, public private partnerships in infrastructure creation and maintenance and service delivery, decentralisation, devolution and deconcentration of expertise and accountabilities within government departments, and contractual relationships between government and community providers, among others (Olukoshi, 2005).

The aim of this chapter is to present the changes that have occurred in the political landscape that have a bearing on the institutional arrangement of public service provision and in turn on solid waste management at the municipal level. Discussions are made alongside three categories of actors (the state-municipalities, the private sector and the civil society). Regional organisations are also seen to play an important role in municipal service provision. The changes are detailed in section 2.2. and they lead to the theoretical discussions in section 2.3 that eventually guide the build-up of the conceptual frameworks on which the study is based. Section 2.4 therefore presents the conceptual models which is followed by a section on the methodology.

2.2 Dimensions of change

2.2.1 *The place of local government*

At independence, that is 1961 for Tanzania, 1962 for Uganda and 1963 for Kenya, all three countries inherited rather strong local governments (LGs) with substantial responsibilities for services. Yet even with the existence of these decentralised systems for all the three countries, the period immediately after independence was one of centralisation. This was justified on the grounds that central policymaking and planning were necessary to bring about the rapid economic and social transformation required. Some of the existing decentralised systems were thereafter abolished while others had their powers reduced substantially. A brief look at each of the three countries just after independence reveals the following.

Uganda

At independence, Uganda was bequeathed with fairly autonomous and well developed local administrations. They were, however, largely ineffective and inefficient, and weighed down by sectarian (clan, ethnic and religious-political factionalism)[3] (Mutebi, 2008). The post-colonial government therefore inherited a local government system already beset with problems. The search for solutions during the Milton Obote regime that began in 1962, was undermined by the government's greater enthusiasm for the elimination of opposition at all levels of government and stifling dissent within the ruling party itself. Local government was thus subjected to the will of the ruling party. The Local Administrations (Amendment) Act 1962 marked the beginning of a series of legislative measures that stripped local administrations of most of their powers and autonomy. The government justified these measures on the grounds that district councillors were of low levels of education and lacked experience. Reacting to the local conflicts its reforms had re-ignited, the government passed yet another Act, the Local Administration (Amendment) (No.2) Act 1963. All secretaries-general, finance secretaries and chairmen of district councils were dismissed. These and other changes with the passing of the Local Administrations Act, 1967 essentially put local governments at the mercy of the centre and provided the central government with the opportunity to influence the functioning of local administrations. Centralisation, however, had many unintended consequences. Civil servants and local administrations continued to perform poorly. Political factionalism, stocked by the ruling party's search for dominance, crippled local administration's capacity, even to collect revenue, which in turn rendered them incapable of performing their functions. But the regime of Idi Amin, beginning in 1971, went even further and abolished district councils to reduce costs of administration next to other objectives. Domination by the centre went on until 1986 when National Resistance Movement (NRM) opted for devolution.

---

[3] At independence Uganda had a multi-party political system with three main parties: the Uganda People's Congress (UPC); the Democratic Party (DP); and Kabaka Yekka (KY). Each party represented particular ethno-religious and regional interests. Broadly, the DP was allied with the Catholic Church. The UPC had strong connections with the Protestant Church, although it worked closely with the Muslim community as well. The KY was a Protestant-leaning royalist party.

Under the Movement system, the Local Council (LC) system was originally proposed as democratic organs of the people in order to establish effective, viable and representative local authorities. Since then a number of important steps have been taken. In 1993 a first thirteen districts were decentralised, and they were given the authority to retain a proportion of the locally generated revenues. A new national Constitution adopted in 1995 clearly stipulated the principles and structures of the LC system. Following this new Constitution, the Local Government Act 1997 was enacted. The Act was made to give full effect to the decentralisation policy. The objective of decentralisation is to ensure good governance and democratic participation in the decision making process. Today there are different levels of local councils, five in total. Local Council three (LC3) is charged with service delivery at the municipal level including solid waste management.

Tanzania

In Tanzania, following colonisation, the British local government model was adopted. In the post-colonial era the local government system, starved of resources, was unable to deliver adequate services to the people (because of among other reasons gross mismanagement of funds collected and granted by the Central Government; little or no capacity to maintain and run a lot of infrastructures constructed by the Central Government and a general lack of qualified personnel). In 1972 local governments were abolished in favour of a more centralised system of government. Central government and line ministries were put in charge of the administration of basic government services at the local level, including primary education and health care. However, the delivery of public services actually deteriorated under this system of deconcentration, and local governments were re-introduced by the Local Government Acts of 1982.

While Local Government Authorities, (LGAs), were technically reintroduced in mainland Tanzania in 1984, the system was a top-down model and local governments were tightly constrained by the central government bureaucracy. This system also failed to yield the desired improvements in the delivery of local services, while stifling local democracy, and, by the early 1990s it had become evident that a fundamental reform of the system was imperative.

By the 1980's Tanzania was the world's second poorest country in GDP per capita terms which certainly affected service delivery at the local level and in 1981, Tanzania tried its own Structural Adjustment Programme (SAP) to address this challenge. For most of the developing countries, the three East African countries included, SAPs were intended to reduce the size and the reach of the state. But for Tanzania in particular, the SAPs brought with them an increase in foreign aid inflows while domestic savings performance deteriorated, making Tanzania more dependent on inflows of foreign aid, raising the question of the sustainability of Tanzania's economic reform program.

During the early 1990s, a Civil Service Reform Program was launched, consisting of six components, including one on Local Government Reform. This component was aimed at decentralising government functions, responsibilities and resources to LGAs and strengthening the capacity of local authorities. Reform of the local government system was initiated in 1996 through a National Conference seeking to move 'Towards a Shared Vision for Local Government in Tanzania'. This vision was subsequently summarised in the Local Government Reform Agenda, and, in October 1998, endorsed by the Government in its *Policy Paper on Local Government Reform* (Mmari, 2005). The overall objective of the policy was to improve the delivery of services to the public, and the main strategy for doing so was decentralisation through devolution, which entails the transfer of powers, functional responsibilities and resources from central government to local government authorities. Today, the devolved structure in municipalities goes down to ward levels with Ward Development Committees considering peoples' welfare in the wards. There are also street leaders in every street to assist on governance issues. Solid waste management in certain municipalities is divided into zones.

Kenya

At the time of independence, the Government of the ruling party was totally opposed to decentralisation, arguing that the structure was unworkable and politically inappropriate – a tactic of divide and rule by the British. And therefore when the ruling party took full control, the central Government began the process of bringing the local state apparatus under its control. In 1964, a parallel administrative system to the LG, the Provincial Administration (PA), was transferred to the office of the president thus strengthening the position of the PA and enforcing its parallel stand further. Meanwhile the LG was not relieved of the expanded responsibilities that they had acquired prior to independence. They had neither the human nor the financial resources necessary to undertake these responsibilities. They were thus forced into a position in which they were bound to perform badly and thus giving the Central government the justification to increase its control. Steffensen *et al.* (2004) explain that the 'Transfer of Functions Act' in 1969 reduced the powers of LG substantially. In 1974 the Graduated Personal Tax (GPT) was abolished and replaced with a centrally controlled sales tax as a source of revenue. This tax had no clear relation with the local tax base as did the GPT. In 1983, the Kenya's District Focus for Rural Development system, was introduced and this incorporated representatives of lower tier elected local government councils, but was essentially a deconcentration of central ministries tightly controlled by officials from the regime. Researchers report that local MPs in collaboration with Presidentially appointed District Commissioners made decisions on local development and resource allocations routinely on the basis of political patronage and access to centrally controlled networks (Crook, 2003).

Since the mid-1990s the Government initiated an incremental reform of LGs that, foremost, focused on improving the fiscal aspects of LGs but until the year 2010 this was done without substantial legal reforms. Generally the government has, until recently, mainly pursued a policy of deconcentration, with only a marginal role for LGs including regarding solid waste management. The new constitution, passed in 2010, makes provisions for substantial devolution of powers to elected LGs. It provides for the creation of 47 counties based on the boundaries of administrative districts dating to 1992. Section 176 of the constitutions states that there shall be a county government for each county and that each county government shall decentralise its functions and the provision of its services to the extent that it is efficient and practicable to do so. 15% of the national revenue is to be sent directly to these counties each financial year and an equalisation fund whereby 0.5% of the annual revenue is given to marginalised communities.

It is evident that in East Africa decentralisation to the municipal/local level has been important on the development agenda for much of the post-independence period (Conyers, 2007). Though these decentralisation efforts were typically politically motivated, they have profound impacts on other aspects, such as governance in the public sector, including public services delivery like solid waste management. This history clearly shows the strive between central and local governments. Until the 1990s for the case of Uganda and Tanzania and 2010 for Kenya, the local authorities were just playing what Davey (1996) refers to as residual roles.

2.2.2 Privatising municipal services

The process of privatisation of municipal services is to varying degrees relatively recent in practically all three countries under study. None of these countries has any municipal service that is completely privatised as yet. The services most experimented with so far are solid waste management and water supply, but the former more than the latter. The following sections depict the trend towards privatisation in the respective countries.

Uganda

The 1960s and 1970s saw the nationalisation of much of the Ugandan economy. The expulsion of the Asian community in 1970 accelerated this trend with many of the departed Asians' properties falling under the Government's control. The present government, which came into power in 1986, was quick to adopt a policy to reverse this trend and promote the private sector's involvement in the economy. The private sector development at this stage was largely a response to the demand for services and benefitted only those who could afford to pay for them. Schools, health clinics, garbage collection companies, recreation facilities and cleaning companies sprung up and that trend is continuing to meet the demand throughout the country. For obvious reasons this demand is higher in high density urban areas where a concentration of markets for these services exists than in less populated rural areas. Urban authorities have supported this trend by providing trading licenses and permits for operation to private entrepreneurs.

The process of privatisation at national level has affected the means through which urban councils acquire resources and run their affairs. At the same time the modes through which their constituents can expect to receive the services traditionally provided by local councils, are changing. The enactment of the Local Government Act 1997 gives urban authorities autonomy over their financial and planning matters. All urban councils now have the power to contract out services to the private sector. They are however, still obliged to establish, prescribe, and control these services and administer the relevant forms.

Although there is no distinct policy on privatisation of municipal services in Uganda, the process of privatisation for municipal services has borrowed from the existing parastatal privatisation policies and Acts. Municipal Councils are considering comprehensive privatisation programs as a means to reinforce and enhance their ability to govern and increase the level and quality of services to their constituents.

Tanzania

After independence, service provision was mainly confined to central government and local authorities through agencies. In a way there was some form of privatisation as service users were obliged to pay taxes, licenses, fees, etc., which contributed directly or indirectly to financing different services. The practice was changed after the Arusha Declaration in 1967 which among other things pronounced the United Republic of Tanzania as a socialist state and advocated for widening the role of the public sector at the expense of the private sector. Thus after 1967, the role of central and local government institutions in service provision was widened and that of the private sector, if anything, ended. The 1982 Local Government Act came into effect at a time when private sector participation in service delivery was at its minimum, as it was only in public transport in the form of town buses that the private sector was involved. New attempts to encourage the private sector to participate in the provision of municipal services were made by the central government. For instance the Trade Liberalisation Policy of 1984 was formulated. However, it was not until the late 1980s and early 1990s that the private sector involvement in service delivery became more noticeable. Given the crisis in service provision which developed out of the pre-1980s policies, various areas attracted the private sector in service provision among which was solid and liquid wastes; privatisation in solid and liquid waste collection and disposal was slowly taking shape in both small and big towns. Among the municipalities where privatisation has been tried the mode has been through contractual agreements between the urban local authority and the contractor taking over the new role, leasing services formerly provided by the authority to the private operator and the introduction of user charges. However problems can be noted for instance in the quality of service offered as well as in the low level of service charges declared to have been collected. The Tanzanian government is yet to formulate an overall policy on privatisation of municipal services provision. Thus, the privatisation initiatives are mainly based on other factors, including pressure from users and the private sector response to exploit the gap between what the public sector offers and the residents demand. In the absence of a appropriate policy, local councils wishing to privatise the provision of services use the existing legislation particularly the Local Government Acts, no. 7 and no. 8 of 1982.

Kenya

At independence, the desire to accelerate economic and social development, to increase citizens participation in the economy and to promote indigenous enterprises led to the establishment of state-owned enterprises. But comprehensive reviews of the public enterprises performance carried out in 1979 and 1982 concluded that the productivity of state corporations was quite low. Following this low performance and in response to the Structural Adjustment Programme, the government instituted some economic policy reforms which included privatisation of public enterprises. Although privatisation of municipal services was not clearly identified as such, Moyo *et al.* (1998) record that the government has been moving towards that direction albeit in a 'stop-go' nature, given the failure of the local government system in the delivery of municipal services. In practice, the provision of municipal services by the private sector has been going on for a long time though informally and without the recognition from local councils. Some of the privatised services include public transport, road maintenance, health, education services and recreational facilities.

Despite the fact that privatisation of municipal services is not new in Kenya, comprehensive policy guidelines are still lacking. However, municipal authorities can and are taking advantage of policy objectives for privatisation process outlined in several documents such as '*Economic reforms for 1996-1998: the policy framework paper (1996)'* and the Eighth National Development Plan and more recently the Economic Recovery Strategy for Health and Employment creation of 2003-2007 and now the Vision 2030. The Local Government Act Cap 265 also allows the municipal authorities to enter into contracts

Overall, for the three countries, there is as yet an absence of clear policies on privatisation of municipal services as well as an absence of appropriate legislation to support privatisation of municipal services. Both central and local (municipal) government authorities are understandably ambivalent about the necessity to privatise certain municipal services. In some cases, there is some residual resistance or lack of enthusiasm on the part of local/municipal authorities to share responsibilities with private sector enterprises, let alone completely give up these responsibilities, notwithstanding their limited capacities to deliver these services themselves. Moyo *et al.* (1998) record that privatisation of some municipal services is seen by them as giving up power, authority and control and municipal authorities and their officials do not find easy to do so willingly. Where real privatisation of municipal services has actually taken place in elements of solid waste management, it has largely been done selectively and piecemeal, serving or operating effectively mainly in the middle- and higher income residential neighbourhoods that can afford to pay for these services. This process left a large proportion of the poor and low-income neighbourhoods, where residents cannot afford to pay for the privatised services, unserved. This has raised issues of equity and social integration and remains an unresolved public policy issue, that is how to provide such services to segments of the population that cannot afford to pay for private sector provision. Noted though is that for all the three countries, motivation for privatisation of municipal services by councils and the national governments range from increasing efficiency to actually developing the private sector, as shown in Table 2.1.

Table 2.1. Privatisation of municipal services: a comparison of stated motives (Moyo et al.*, 1998).*

City/municipality	Increase efficiency	Reduce fiscal burden	Adopt innovation and new technology	Broaden ownership	Increase revenue collection	Develop private sector
Kenya (general)		√		√		√
Nairobi		√	√			√
Mombasa	√	√	√	√	√	√
Eldoret	√	√	√	√	√	
Uganda (general)	√	√		√		√
Kampala	√	√			√	√
Jinja	√	√		√	√	√
Tanzania (general)		√	√	√		√
Dar-es-Salaam	√		√		√	√

2.2.3 Role of civil society

Malunga (2006) gives the definition of civil society organisations as comprising organisations of citizens that come together to pursue interests and purposes for the good of all. This definition includes NGOs, community groups, labour unions, professional associations, faith-based organisations and parts of the media and academia and they may operate at all levels from grassroots at village and community levels to national and international levels. These Civil Society Organisations (CSOs) have been instrumental in municipal service provision and the different roles that they have played over time are described below.

Uganda

The current characteristics of Uganda's civil society find their roots in the country's experiences during the past eight decades, going back to the colonial era, when the State was the main provider of social services within the overall design of an export-oriented economy based on small-holder agricultural production. DENIVA (2006) records that a measured (though highly regulated) development of CSOs was then encouraged, with CSOs primarily consisting of co-operatives of export crop growers and trade unions/associations, as well as mission-established hospitals and educational establishments, and other charitable institutions. These trends went on until the time of independence. After independence, the peasant cooperative societies and trade unions were taken over by the Government and bureaucratised. Consequently, the demarcation between 'civil society' and 'government' started to become blurred. Mission schools were integrated within the state education system; political parties were eventually banned and other forms of political dissent, often associated with the traditional kingdoms, curtailed. CSOs were henceforth confined to the more 'traditional fields' (charity, health delivery) and sustained in doing so by the earlier interventions of 'charity'-oriented international NGOs. The National Resistance Movement government took over power in 1986 and inherited a near collapsed economy. This was then followed by a period of reconstruction and relative freedom that provided space for the emergence of indigenous CSOs. Simultaneously, the relative peace that prevailed in many parts of Uganda after 1986 encouraged people to build their own CBOs, including farmers self-help groups and many other types of voluntary associations. The 1990s also witnessed a very rapid growth in the numbers of CSOs because many donors preferred to channel their financial support to them, as they were considered less corrupt and more efficient.

Today, the legal environment is judged to be not as enabling as it could, given the cumbersome registration procedures for CSOs. The government also shows a hesitant attitude on what constitutes allowable advocacy activities for CSOs, especially when they 'stray' into what it considers the political arena. Nevertheless, this relationship is changing on several fronts (*ibid.*). Both parties increasingly see advocacy work as legitimate for CSOs to engage in, a move supported by donors. Moreover the government is opening avenues, especially at district level, for CSOs to contract service delivery from its own coffers. It is common today to see CSOs awarded contracts even for solid waste management.

Tanzania

The origins of the modern kind of civil society in Tanzania can be traced back at least to the beginning of British colonial rule in the 1920s and this went on even to the time of independence with the civil movement spearheading the struggle for independence. However, after multi-party elections in 1965, the constitution was changed in a manner that allowed the establishment of a one-party system. Nyerere's socialist one-party rule inherited most of the laws and institutions from the colonial period. The state also controlled the legal system, which was used to control and coerce the citizens rather than to guarantee rights for them. This political environment was very restrictive for non-governmental associations and made it virtually impossible for civil society to organise independently. Only religious groups, charity organisations and relief foundations were allowed to operate, because their activities were not considered political. (Lange *et al.*, 2000; Haapanen, 2007). During the 1980s, structural adjustment programs led to increased funding for civil society organisations. Foreign donors' funding strategies emphasised the strengthening of the 'third sector' instead of state institutions, which were seen as inefficient and corrupt. After Nyerere's retirement in 1985, and along with the steps toward multi-party democracy in the beginning of 1990s, CSOs were given new space and they were now seen also by the state as important for community development. The numbers of CSOs started to rise rapidly. The legal framework for CSOs has evolved during the multi-party era, and especially NGOs' position is now established and guaranteed by law. Today, there is much diversity among civil society actors in Tanzania. A major part of Tanzanian civil society consists of informal groups and small community based organisations (CBOs), professional associations and trade unions, as well as numerous faith based organisations (FBOs). In terms of numbers, local CBOs and informal groups may be the main actors, but there are no exact numbers available, because a substantial part of these groups are not officially registered. Usually, these smaller organisations operate at grassroots level, particularly with the poor, disadvantaged and marginalised people, in helping to improve their social situation and living conditions. Generally, there are concerns about the capacity of local CSOs to take up and utilise the political space that enables them to have influence on governmental decision-making.

Kenya

Civil society, and in particular the NGO sector, has undergone fundamental changes in post-colonial Kenya. As in the period of colonialism, the state has played a central role in defining the direction of the voluntary sector especially as relates to its vibrancy. But one thing that is also certain is the fact that CSOs have increasingly taken on important tasks in society, and have in their own different ways equally influenced the nature and character of the post-colonial state. Accepting the enormity of the development problem in the 1960 and 1970s, the Government actively encouraged self-help efforts in the generation of development. This was marked in the beginning by the important role given *harambee* (Swahili word for self-help) in Kenya's development process. But *harambee* had mixed outcomes which negatively affected its credibility. In addition to this form of voluntary self-help, many NGOs, both formal and community-based organisations, were rapidly expanding their activities and numbers in the country. And most of them especially the NGOs in the period between 1964-1978 were largely involved in development activities as opposed to political activism. A number of them later got involved in political activism following the 1982 constitutional change that made the state a de jure one-party system and thus legally barring the formation of opposition political parties that could provide an alternative to KANU (the ruling party then). With multi-party arrangement and liberalisation of the national economy in the 1990s together with a new political regime in 2002, however, more and more CSOs are seen taking up roles even in the public service provision. CSOs are today engaged in government budget processes in order to affect policy choices and make public budgeting more open and accountable. There is the local Authority Transfer Fund to local authorities whose conditions require the involvement of CSOs in decision making at the municipal level. Due to their diversity in activities, the small CSOs -in terms of expenditures and the number of people involved- (CBOs and self-help groups) remain only dimly understood, making it difficult to determine what their capabilities really are (Karuti *et al.*, 2007). Some operate without registration, while others operate after informing the provincial administration, but without documentary evidence of existence.

Overall, a major trend in the CSO sector over the last decade for the three countries, but more so for Uganda and Tanzania, has been the awarding by governments of contracts to CSOs for the delivery of services (Clayton *et al.* 2000). Clayton goes on to write that while the CSO sector may have grown enormously in terms of size, its independence has been compromised and this has not improved the capacity of CSOs to provide an alternative development agenda for donors and governments. A study by Robinson and White (1997) argues that while CSOs play an important role, especially where state provision is weak and the private sector caters to the better off, there are a number of common deficiencies with the services provided by the CSO sector. These include: limited coverage; variable quality; amateurish approach; high staff turnover; lack of effective management systems; poor cost effectiveness; lack of co-ordination; and poor sustainability due to dependence on external assistance. Even with these limitations, studies have shown that CSOs are needed at the interface between people and decentralised state bodies, the implementation level, to represent the interests of the poor and facilitate their participation in decision making. Furthermore, community involvement allows for a design that responds, from the beginning, to local needs and Karanja (2005) concludes that mobilising communities makes it possible to achieve more with limited finances.

2.2.4 Regional integration

Today service provision at the municipal level is characterised by regional actors who according to Olukoshi (2005) play an important, even at some conjunctures determinant role in shaping outcomes and therefore are critical to the definition of the process of change. Regional integration in East Africa can be traced to the 60's.

The decade of independence in the 1960s witnessed a peaceful and united East and the Horn of Africa. United in the common agenda of securing regional peace and tranquillity; combating environmental threats and disasters, and investing in trans regional infrastructural development, the region excelled in education, communication and cooperation. But soon after, this cooperation was shaken by a number of political upheavals ranging from the Sudanese civil war of 1983, the 1994 Rwanda genocide and most recently the Somalia mayhem which still rages on. The repercussions of these disturbances affected neighbouring countries and at some point only Kenya, Tanzania and Djibouti were islands of peace in the sea of turbulence that marked the political landscape of the East African region. Today the East African countries of Kenya, Uganda, Tanzania, Rwanda and Burundi realise that they have a responsibility towards each other, and that they have many challenges in common, some of which are cross-border in nature. These countries now see considerable benefits of close regional cooperation, political understanding, economic and social prosperity. This new resolve to address issues of common relevance in a cooperative manner is reflected in concrete progress being made on the ground. Maruping (2005) records that these countries are today all members of one or more regional or sub-regional arrangements that seek to promote economic coordination, cooperation or integration among the member countries. Mackie *et al.* (2010) report on the various regional organisations whose agendas have been either economic, political or environmental. These include:

1. The Common Market for Eastern and Southern Africa (COMESA) integration which was set out in the Preferential Trade Agreement Treaty of 1993, and is focused on the integration of markets through the removal of tariff and non-tariff barriers to trade and investment to achieve full economic cooperation through a gradual process of creating a free trading zone, the establishment of a common market and ultimately an economic community.

2. The Intergovernmental Authority on Development (IGAD) in Eastern Africa which was created in 1996 to supersede the Intergovernmental Authority on Drought and Development (IGADD), founded in 1986. The main objective of this organisation is to foster economic and diplomatic cooperation between its member states with the aim of increasing food security, environmental protection, economic integration and peace and security in the region.

3. The East African Community (EAC) is an intergovernmental organisation, regrouping five countries from East Africa, created in 1967 to supersede the East African Common Services Organisation. This entity aims at strengthening the ties between member states through a common market, a common customs tariff and a range of public services so as to achieve balanced economic growth within the region. In the last ten years, the EAC has launched several projects, at the regional/sectoral level, in support of deeper integration within the region. These include, amongst others: the single tourist visa programme to facilitate free movement of tourists in the region so as to make the region a more attractive and competitive destination for middle class and high class families; the Lake Victoria Development Programme to coordinate and promote investment/information sharing among various stakeholders in the region as a way to transform the Lake Victoria Basin into a real economic growth zone. Other important outcomes of EAC in the Lake Victoria Basin include the formation of sub-regional organisations like the Lake Victoria Basin Commission which contributes to municipal service provision in diverse ways.

These regional initiatives have to deal with challenges such as overlapping membership which weakens integration, inconsistencies between national policies and regional integration programs, lack of continuity and sustainability of regional programs, limitations of the regional organisations' financial, technical and institutional capacities and even the definition of what constitutes a regional program.

To this end, we see a history in urban governance marked with different actors, each playing a role in public (municipal) service provision and advertently pointing to different institutional arrangements that go beyond the state as the traditional public service provider. This setting generates interest in looking at a number of theoretical perspectives that argue for different institutional arrangements in order to bring to perspective the modernised mixtures argument introduced in chapter one.

2.3 Theoretical perspectives

The history just presented depicts an institutional arrangement for which the central government in all the three states has had a powerful leading position as far as municipal level is concerned. But we also see non-state actors involved in one way or another from local to international level. A number of authors have argued for and against different institutional arrangements and this study picks on some of these arguments. In particular the theoretical perspectives of centralisation vs. decentralisation; developmental state vs. network governance; and multi-level governance to discuss how these arrangements at municipal level affect decentralised governance and environmental service provision (SWM).

2.3.1 Centralisation versus decentralisation

While it is not the intent of this study to emphasise the merits of decentralisation (i.e. as compared to centralisation), the political arena in East Africa is characterised by distinct centralised and decentralised features making it necessary to take a closer look at them. The different positions taken in the centralisation versus decentralisation debate give insights into the effects that a centralised system would have on environmental service provision as the study seeks an optimal institutional arrangement for the decentralised public services.

A number of authors, among them Crook and Manor (1998), Oyugi (2000), Fjeldstad (2001), Crook (2003), Mitullah (2004), Conyers (2007), and Robinson (2007) record that decentralisation has done little to improve the quality, quantity and equity of public services. Yet others (Rondinelli *et al.*, 1984; Azfar *et al.,* 1999; De Vries, 2000) report that centralisation does not take local stakeholder knowledge and interests into account. Centralisation imposes a top-down view which often benefits the vested interests as opposed to those with less power in the process, while ultimately it is only at local level that policies can be implemented and decentralisation will ensure effective and probably more cost-effective implementation.

The two schools of thought are looked at here against the following criteria: efficiency, resource availability, participation and equity in public service provision and in turn in solid waste management at the municipal level. In this review we borrow on discussions by Prud'homme (1995), Klugman (1994), De Vries (2000), Wunsch (2001), Jutting *et al.* (2004) and Ahmed *et al.* (2005), who write on decentralisation, centralisation or on both in relation to public policy, service delivery and in some cases with particular reference to Africa.

When it comes to *efficiency,* Klugman explains that the design of public goods and services may be more in accordance with local preferences under a decentralised system (allocative efficiency), but weighing against this are central standards which can ensure a minimum degree of quality and quantity of the services provided. Further, the utilisation of local resources, information and technology may lower costs (technical efficiency), but the existence of economies of scale points in favour of a more centralised approach to service provision. A certain minimal area and, therefore, a certain degree of centralisation seem necessary prerequisites for the provision of some public services, or at least for their cost-efficient provision. She goes on to say that local decisions which rely solely upon local information may ignore the potential for externalities and economies of scale, which in turn creates wider costs for the society. Yet dis-economies of scale may also exist, where costs rise disproportionately with size.

Looking at *resource availability*, a frequently-cited problem is the lack of capacity at subnational levels of government to exercise the responsibility for public services (Wunsch, 2001). The possible advantage of effectiveness under a decentralised system is disputed, because local governments lack the financial resources to finance the decentralised policies (see De Vries, 2000; Prud'homme, 1995). Furthermore, large public agencies are seen as necessary for the recruitment of qualified and skilled staff. Under a decentralised system, professional staff may be relatively more isolated and lack the frequency of contact with colleagues, than would have occurred under centralisation. This has been said to lead to a lower quality of service provision than would otherwise have been the case. Yet decentralised institutions and policies are associated with innovations which happen because good ideas bubble up from employees, who actually do the work and deal with the local people

From the angle of *participation*, decentralisation might be favoured because it is deemed to increase effective democratisation. This is the basis of empowerment for local communities. To proponents of decentralisation, the centralisation of policy strategies at national level is inimical to citizenship because it deprives local civic forums of weighty matters to deliberate about. De Vries (2000) who presents the arguments in favour of both centralisation and decentralisation of public policy making, points out that the use of local people for the provision of local services would render these services cheaper and more effective. An example given is the low cost Orangi sanitation pilot project in Karachi that involved the community. De Vries goes on to write that decentralisation can cut through red tape and may increase officials' knowledge of and sensitivity to local problems. This may result in better penetration of national policies to remote communities, greater representation for various religious, ethnic and tribal groups in the policy process, and greater administrative capability at the local level. Decentralisation can provide a structure in which local projects can be coordinated, civic participation enhanced, and entrenched local elites, who are often unsympathetic to national development policies, neutralised. It may result in a flexible, innovative and creative administration, and be more effective in its implementation, because of simplified monitoring and evaluation. On the other hand, decentralisation is also reported to increase the possibility of political capture within these lower governmental tiers (Jutting *et al.*, 2004; Shah *et al.*, 2005; Wunsch, 2001).

As relates to *equity*, Klugman (1994) explains that while decentralisation can promote equity among different groups within a region, through increased local public expenditure and the wider provision of public goods and services, there is a risk that disparities between districts may worsen due to different tax bases and inadequate fiscal equalisation. Therefore there may be a need for greater decentralisation in some cases or centralisation in others so that the central government at least retains a strong redistributive role.

For some an absolute choice for centralised or decentralised options is necessary, but like Hegger (2007) argues, none of the two extremes is completely preferable over the other. What is coming out clearly at this point is that the arguments used for or against centralisation can at other moments also be used to make opposite claims. The features of decentralisation are likely to differ for different countries and depend on the environmental service in question. Decentralised responsibilities in some areas is more sensible than in others especially for issues having national causes and consequences, that is, services that are provided at the national level. A number of other aspects also play a crucial role when looking at decentralisation for instance, the dependency on central grants, autonomy of local government in the policy area, possibility for the public to participate among others. This study therefore builds on the theoretical arguments posited above by adapting them to specific existing situation because as De Vries concludes, it cannot be pointed out often enough that an optimal institutional arrangement fits the specific situation in a specific area in a specific country given the specific problems at stake.

2.3.2 Developmental state versus network governance

This debate takes the study a step further from solely focusing on the government as an institution (both local and national) to giving attention to the non-state actors who are increasingly taking an important place in public service provision. Emphasis is thus on the adequacies in the role of African state in urban environmental infrastructure development and public service provision in East Africa and the presence of non-state actors therein. Questions regarding the adequacy of the structural arrangements of public authorities are raised. Therefore this study looks at the arguments for 'neo-developmental state' – a concept brought forth by Oosterveer (2009), which suggests a renewed and active role for governments in promoting development and ensuring adequate service provision also for the poor, and, on the other hand, the 'network approach to governance' which acknowledges limitations to the state's capacity and suggests a further involvement of other societal actors in the governance of urban environmental infrastructures and services.

The developmental state is different from the traditional authoritarian state because its legitimacy derives from its achievements and not from the way it came to power (Johnson, 1999). Proponents of the developmental state point out that states should foster economic development and avoid being captured by particular interest groups (Krieckhaus, 2002; Menocal and Fritz, 2006). Thus, a developmental state is broadly understood as one that shows a clear commitment to a national development agenda, that has solid capacity and reach, and that seeks to provide economic growth as well as poverty reduction and the provision of public services (Menocal and Fritz, 2006). Referring to Africa specifically, Mkandawire (1998) gives the definition of a developmental state as a state whose ideological underpinnings are developmental and that seriously attempts to deploy its administrative and political resources to the task of economic development. The theory of developmental states, as explained by Krieckhaus, argues that states with a combination of high bureaucratic capacity and significant autonomy from society can successfully allocate financial resources to strategic industrial sectors and thereby generate rapid industrial advances. Developmental states are seen to be able to generate massive savings within the public sector and use these resources to finance high levels of investment and growth. Under this school of thought, civil servants are more professional and more detached from powerful rent-seeking groups attempting to influence them (Menocal and Fritz, 2006). Oosterveer (2009), argues for a revised understanding of the developmental state, as this has evolved from its initial form in the 1960s and '70s into a 'neo-developmental' state in the new millennium. Whereas the original developmental state was oriented towards monopolising the process of economic and societal growth, its renewed form acknowledges the role of other sectors in society (market, civil society) but insists on the essential role of governments in protecting the interests of the poor. Oosterveer gives an elaborate theoretical discussion of the role of the neo-developmental state in urban environmental infrastructure and service provision in East Africa. He reports that under the view of the neo-developmental state, privatisation is not excluded but should be firmly controlled (against corruption and underperformance) and not be limited to contracting large (foreign) companies, but include smaller local companies and NGOs/CBOs. Oosterveer further records that proponents of the neo-developmental view claim that it is only through active government interventions that access of the poor to environmental infrastructures can be secured, as their economic and political power is too limited to realise this in another way. Stronger national and municipal governments are considered better able to bridge the gap that very often exists between the formal laws and regulations and their limited implementation in practice. Effective market economies are essential but require functioning and capable state and societal institutions in order to operate and grow.

Critics claim that assigning an active developmental role especially to the contemporary African state is anachronistic (Kütting, 2004 in Oosterveer, 2009). A neo-developmental state demands strong governmental organisations, which becomes increasingly unlikely under the present economic and international conditions (Callaghy, 1993; Lewis, 1996). Instead of strengthening the African states' capacity to intervene they suggest to limit their role even further and rely more on other societal actors through network governance.

Network governance involves the collaboration between various institutions and structures of authority to allocate resources and to coordinate and control joint actions across the network as a whole. A number of authors have contributed to the discussion of network governance. The works of Jones *et al.* (1997), Provan and Milward (2001), Provan and Kenis (2007) in their own different ways, assess the effectiveness of networks. Provan and Kenis (2007) examine the functioning and governance of networks, thereby distinguishing between organisational and network governance. They even go further and present different forms of network governance. They ultimately present what they call tensions in network governance tied to legitimacy, stability and efficiency in relation to inclusive decision making. Jones *et al.* (1997) do not refer explicitly or exclusively to a public sector context but focus on a general theory of network governance. Using transaction cost economies and social network theory, they provide an assessment of alternative forms of governance. They identify conditions for network governance and explore why networks, rather than markets or hierarchies, are employed. Provan and Milward (2001) focus on the definition of network performance indicators by identifying three relevant levels of analysis: community, network, and organisation/participant. At each level the authors propose a set of criteria to measure effectiveness and conclude that overall effectiveness of the network in service delivery will ultimately be judged by community level stakeholders. Oosterveer (2009) records that a network approach permits inter- organisational interactions of exchange, concerted action, and joint production in a more or less formal manner. The composition of such networks vary from domain to domain, but they are likely to consist of government agencies at different levels, key legislators, pressure groups, relevant private companies and civil society organisations such as

NGOs and CBOs. Network governance arrangements intend to achieve their objectives through the combined efforts of these different sets of actors, but their respective roles and responsibilities remain distinct without the state being the sole locus of authority. He goes on to write that these non-state actors start developing their own sets of rules or standards to fill 'institutional voids' where rules to guide behaviour are needed but not provided by the state. The national government remains an important political actor and a point of orientation for citizens but must compete with others and can no longer assume a monopoly on legitimate interventions. Oosterveer concludes that the network approach to governance seems more promising to deliver urban environmental services to the poor in East Africa than the neo-developmental approach.

Network-based views are criticised for the lack of legitimacy of the non-state societal actors involved. Unlike state-based regulators, whose actions can be legitimised via formal, democratic procedures and supported by law, non-state actors cannot rely on legal authority to motivate people, nor derive legitimacy from their position in a wider official order (Provan and Kenis, 2007; Oosterveer, 2009). There is also the question regarding the balance between the need for administrative efficiency in governing infrastructures and services and the need for stakeholder involvement, through inclusive decision making. The more participants are involved in the network decision process, the more time consuming and resource intensive that process will tend to be (Provan and Kenis, 2007). Furthermore while networks are discussed as adaptable, flexible forms that are 'light on their feet', the issue of stability necessary for developing consistent responses to stakeholders and for efficient network management over time come to the fore, especially where the networks are not temporary outfits and where there is shared participant governance as opposed to lead-organisation governance (*ibid.*).

The two schools of thought provide valuable (theoretical) insights into what service provision can gain from either neo-developmental state or network governance each on its own. The idea of neo-developmental state in Africa has been criticised and even dismissed by some. While the East African countries are not on the same footing with their East Asian counterparts from where the developmental state model emerged, we can certainly talk of a neo-developmental state in East Africa. Mkandawire (1998) writes that the answer lies in institutional arrangements and he goes on to say that the way forward does not lie in the wholesale neglect of existing capacities in the quest for 'new' ones, but in utilising, retooling and reinvigorating existing capacities. Yet it is becoming increasingly clear that development cannot be left to the state alone, It has to involve the participation of the people concerned. Network governance arrangements create room for effective local communities' participation in developing their own preferred approach while harmonising with other stakeholders and other levels of governance. Oosterveer (2009) argues that this form of governance could assure that services reach the urban poor. Therefore there is need to look at the two schools of thought and build on the strength of each. This study provides empirical evidence to build on the arguments presented.

2.3.3 Multi-level governance

To continue the discussion on state and network governance, the study also looks at multi-level governance (Harlow and Rawlings, 2006). A defining characteristic of multi-level governance

(MLG) is the pooling and sharing of authority and influence between actors and across different levels (Develtere *et al.*, 2005).

Marks and Hooghe (2003) explain that, in recent decades, centralised authority has given way to new forms of governing. Formal authority has been dispersed from central states both up to supranational institutions and down to regional and local governments. At the same time, public/ private networks of diverse kinds have multiplied at the local and international levels. Marks and Hooghe (2004), explain that the diffusion of authority in new political forms has led to a profusion of new terms: multi-level governance, multi-tiered governance, polycentric governance, multi-perspective governance, FOCJ (functional, overlapping, competing jurisdictions), fragmentation (or spheres of authority), and consortia and condominiums, to name but a few. This dispersion of governance across multiple jurisdictions is seen as both more efficient than, and normatively superior to, central state monopoly. It is also seen as more flexible than concentration of governance in one jurisdiction.

Literature on MLG claims that governance must operate at multiple scales in order to capture variations in the territorial reach of policy externalities. Because externalities arising from the provision of public goods vary immensely - from planet-wide in the case of global warming to local in the case of city services like waste management - so should the scale of governance. Multiple jurisdictions also reflects better the existing heterogeneity of preferences among citizens. However, beyond the presumption that governance has become (and should be) multi-jurisdictional, there is no agreement about how multi-level governance should be organised. Marks and Hooghe distinguish between Type I and Type II MLG:

Type I MLG describes jurisdictions at a limited number of levels. These jurisdictions – international, national, regional, meso, local – are general-purpose. That is to say, they bundle together multiple functions, including a range of policy responsibilities and, in many cases, a court system and representative institutions. The membership boundaries of such jurisdictions do not intersect. This is the case for jurisdictions at any one level, and for jurisdictions across levels. In Type I governance, every citizen is located within a certain set of nested jurisdictions, where there is only one relevant jurisdiction at any particular territorial scale. Territorial jurisdictions are intended to be, and usually are as well, stable for periods of several decades or more, though the allocation of policy competencies across jurisdictional levels is flexible.

Type II MLG is distinctly different. It is composed of specialised jurisdictions. Type II governance is fragmented into functionally specific pieces – say, providing a particular local service, solving a particular common resource problem, selecting a particular software standard, monitoring water quality of a particular river, or adjudicating international trade disputes. The number of such jurisdictions is potentially huge, and the scales at which they operate vary finely. There is no great fixity in their existence. They tend to be lean and flexible – they come and go as demands for governance change. Each public good or service should be provided by the jurisdiction that most effectively internalises its benefits and costs.

Following the above grid that has often been used to research MLG in Europe, Develtere *et al.* (2005), in their study on MLG in Kenya write of the two types 1 and II MLG as vertical and horizontal dimensions respectively. The 'vertical' dimension refers to the linkages between higher and lower levels of government, including their institutional, financial, and informational aspects. Here, local capacity building and incentives for effectiveness at sub-national levels of government are crucial issues for improving the quality and coherence of public policy. The 'horizontal' dimension refers to cooperation arrangements between regions or between municipalities as a means by which to improve the effectiveness of local public service delivery and implementation of development strategies. This study builds on the categorisation above, particularly the horizontal or type II MLG which calls for opening up the political space to non-state actors. Hereby this study appreciates the conclusion from Develtere and colleagues that MLG in the Kenyan context, as a system of pooling and sharing of authority and influence, does not come into being on the basis of a predesigned plan but is instead the result of many different actors, pursuing very divergent agenda, getting interested in a more participatory multi-stakeholder mode of governance that works at different levels in order to satisfy their specific interests.

Type II MLG thus provides a good basis to assess the opportunities of inter-municipal cooperation for SWM and further the role of regional organisation and networks in environmental service and infrastructure provision.

Looking at these theoretical perspectives from the angle of modernised mixtures, the study can at this point argue that it is the intelligent and flexible mixtures of institutional layouts and management systems from opposing paradigms but adapted to the specific local context, that will pass for a good arrangement for SWM at the municipal level. SWM is one of the environmental services considered a public service and mandated to the municipal authorities. Section 2.4 therefore presents conceptual frameworks of analysing the new political configurations at the municipal level that show a mix of institutional layouts and management systems.

2.4 In search of new political configuration for SWM at municipal level under the modernised mixtures approach

There is need to conceptualise new political configurations under the MM framework that take cognizant of the roles played at different government levels by different actors (both state and non-state actors). These actors operate at the local, national and regional levels. The actors perform SWM tasks which are both technical and social and operating along the waste chain either as combined or differentiated waste flows and this translates into different options or combinations of options for institutional arrangements which offer solutions to the problems facing SWM in East African cities. The conceptual models 2.1 and 2.2 below illustrate this (see Figure 2.1 and 2.2).

Solid waste management is the underlying subject of study and some of its elements that are critical for its assessment at this point include waste collection; transportation, treatment/ recycling and disposal. These are tasks under the waste chain and involve a number of actors at different stages who define the different institutional arrangements.

The *municipalities* are the key actors in solid waste management. Their activities are influenced by the *state* and the citizens in the *neighbourhoods/wards* to whom they seek to provide services. Therefore important questions which are part of research *questions 1 and 2*, are whether these municipalities …

- have the tools/resources;
- are organised to perform their functions; and
- perform as expected?

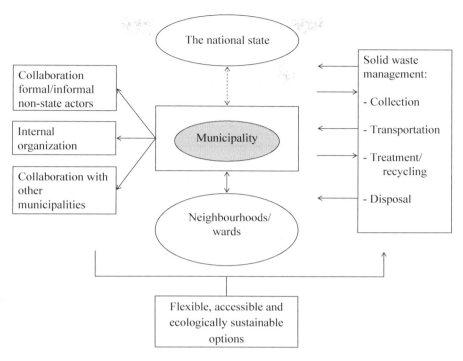

Figure 2.1. SWM at municipal level: actors and relationships.

Answering these questions, allows the research to look at the *internal organisation* aspect of the framework. Other issues that are part of the internal organisation include: the actors involved and their power relations (in this case issues of autonomy), the policy documents and how they have been interpreted and how they have accommodated new definitions of problems or new approaches to solutions (Arts *et al.*, 2006)

Note that in Figure 2.1 the arrow between the state and municipality is dotted because in the study, the municipality is at one point looked at as local government whereas the state is the central government while elsewhere the municipality is considered as constituting the state itself.

It is evident, that there are other non-state actors involved in SWM. Their participation is either formal or informal and to assess this, the study looks at their *collaboration* with each other and with the municipalities (*formal/informal*) where present. Some of the issues assessed are the legitimacy and decision making power: the relationships between the municipalities and the non-state actors, amongst non-state actors themselves and between the non-state actors and the households in the wards/neighbourhoods. The existing payment systems are also looked at.

The small municipal councils which are also cases under study could *collaborate with each other* in their efforts in solid waste management. This can be realised through inter-municipal cooperation within the lake basin. Specific issues looked into for inter-municipal cooperation in the study include the geographical location, institutional organisation, waste collection, transportation, re-use/recycling and disposal and the opportunities presented in line with economies of scale,

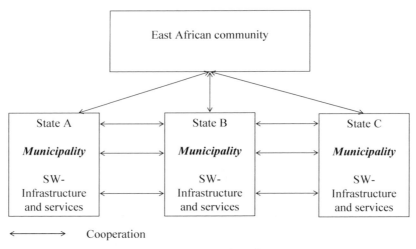

Figure 2.2. Cross border actors and relationships in the Lake Victoria Basin.

equity, subsidiarity, internalising externalities and participation of other public and private entities. This allows the study to answer research *question 3*.

The ultimate options for solid waste management should reflect accessibility, flexibility and ecological and institutional sustainability which are the criteria of the modernised mixtures approach.

SWM can also be done across borders, that is cross-country and to look at this more in detail, conceptual model 2.2 was developed (see Figure 2.2).

Cross-border cooperation has to take into consideration constraining and/or enabling factors at the regional, national and municipal levels and specifically those considered relevant for solid waste infrastructure and service provision. Two regional organisations are assessed for their role in enhancing cross-border cooperation in infrastructure provision by looking at municipal autonomy in decision making within their organisational arrangement and the availability of resources within these organisations. This allows the study to answer research *question 4*.

To operationalise these models, the next section presents the research methodology that aided in the collection of data.

2.5 Operationalising the conceptual frameworks: research methodology

This section presents the overall research methodology but specific details are given under each empirical chapter that follows hereafter.

As mentioned in chapter one, this study seeks to contribute to the improvement of solid waste management in the Lake Victoria Basin. More specifically, attention is focused at the municipal level as shown in the conceptual models above, and the aim is to present options of institutional arrangements under the modernised mixtures approach that would improve infrastructure and service provision. With the three countries in question, it was of necessity to study cases in each of the three and compare them in order to make an informed appraisal that reflects the real life context. This required an in-depth study of each case. In-depth insights were sought with regard to existing institutional arrangements for solid waste management and changes that have occurred over time. To sufficiently do such a study, case study research strategy was used because as explained by Noor (2008) and many other authors, case studies become particularly useful where one needs to understand some particular problem or situation in-depth, and where one can identify cases that are rich in information.

2.5.1 Case study research

According to Yin (1994, 2009), case studies can be exploratory, descriptive and explanatory. De Vaus (2001) gives further categorisations of either single or multiple case; holistic or embedded units of analysis and parallel or sequential case studies. With the theory already in place, research questions defined and the intention to solve a problem – which is solid waste management – then descriptive and explanatory case study designs are more appropriate. This is because the study seeks to show how and why a modernised mixtures approach would be appropriate in realising institutional arrangements that would improve solid waste management.

Case selection

In order to study cases in each of the three countries, three urban centres in Kenya, Uganda and Tanzania were selected as multiple cases for comparative purposes on the basis of:
- diversity of experience, including diversity within one country, and therefore the potential for improving our understanding, identifying good practices and disseminating relevant lessons;
- feasibility of the field research in terms of the researcher's existing level of knowledge and contacts;
- geographical representation of the Lake Victoria Basin;
- existing partnerships, expertise and knowledge of the centres and countries.

The three selected urban centres are Jinja (Uganda), Mwanza (Tanzania) and Kisumu (Kenya). Besides the points listed, in their respective countries the three towns all come just after the capital cities, in terms of urban status, and therefore their institutional position is comparable.

In responding to the first and second research question, the three cases, Jinja, Kisumu and Mwanza municipalities, help to show how the modernised mixtures approach holds for different conditions. The research was able to make a parallel comparative study of the three urban centres, whereby all three were studied at once. The comparative approach brought out contrasts in institutional organisation which would have been lost in a single case study.

The third research question is addressed by looking at embedded units of analysis. With lake Victoria Basin being considered as the main case, three small urban centres also referred to as secondary towns neighbouring each other were picked from the Kenyan side of Lake Victoria and they formed the embedded units. This approach was informed by the findings from research questions one and two which showed clear differences in the organisation and legal provisions of urban authorities in handling solid waste management amongst the three countries, therefore making it more appropriate to consider inter-municipal cooperation on infrastructure provision within one country. The three towns were purposely picked out of 6 municipalities in Kenya[4] because of their geographical proximity in the Lake Victoria basin. The three small urban centres are all within 100 kms of each other.

To respond to the fourth research question, two regional organisations in the Lake Victoria Basin were selected, one categorised as voluntary (Lake Victoria Region Local Authority Cooperation) and another as statutory (Lake Victoria Basin Commission) in order to assess how the roles they play (if any) in solid waste management at municipal level are conditioned by their organisational structure.

2.5.2 Data collection

Since the cases consist of different elements, different methods of data collection were used (see also De Vaus, 2001). A number of issues in the research design required in-depth interviews, hence in these case studies the emphasis was on qualitative methods. Interviews were conducted with city/municipal and town council officials responsible for solid waste management in one way or another to assess the institutional arrangement at their respective council and for the smaller town councils additional information was sought to assess the opportunities for inter-municipal cooperation. Interviews were also made with representatives of non-state actors in solid waste management (private companies, CBOs, solid waste management associations) to assess the relationships that exist amongst themselves, with the councils and with the households/citizens (see Annex 8). Representatives of regional organisations were also interviewed to assess their role in SWM and to establish if there are opportunities for municipal cooperation. Details of each part of the research are further explained under each chapter.

Two stakeholder workshops also informed this study. One held in Kisumu brought stakeholders particularly from the small urban centres to a round table discussion on opportunities for inter-municipal cooperation and more details on this workshop can be found in Chapter five. One of the outcomes of this workshop which is important for the study concerned the actual and potential roles that regional organisations play in SWM. Specific aspects analysed under areas for cooperation included economies of scale, internalising externalities and ensuring equity amongst participating councils which allowed the study to conclude on the flexibility, accessibility and sustainability of inter-municipal cooperation. There was also a Bukoba workshop organised under the auspices of LVRLAC which had a wider spectrum of participants compared to the Kisumu workshop, with participants from Kampala, Mwanza and even Kisumu. The role of regional organisations and networks in SWM was looked into and the specific areas analysed included, the autonomy of participating councils and the availability of resources, reflecting the flexibility of regional arrangements and their institutional sustainability.

The research also entailed looking into legislations, policy documents and other relevant material. This was helpful in discovering the history surrounding the institutional arrangements for public service provision and in understanding the different theoretical perspectives that assisted in the development of the conceptual models in line with the modernised mixtures approach.

---

[4] The Kenyan side of the Lake Victoria Basin has 25 councils: one city Council; 6 municipal councils; 7 town councils and 11 county councils which are rural.

Direct observation of SWM infrastructure was necessary to provide a description of the status quo in the urban centres under study.

To complement the qualitative approach and to answer certain questions, questionnaire-based household surveys were conducted in Jinja, Kisumu and Mwanza with a total of 200 questionnaires administered in each of these three urban centres. A number of 200 questionnaires is considered enough for a first impression of the situation in each town because the study ensured to have the samples randomly spread over the different sections of each town. The number 200 is also the maximum the researcher could afford in terms of time and money.

Sample selection for the household surveys

Mwanza

In Mwanza, there are two districts and each has urban and rural wards. Urban wards were purposively selected because they receive SWM services. Every urban ward receives services either from a CBO or a private company working under a contract so all urban areas irrespective of income levels receive SWM services. For the survey, stratified sampling was used. It was a disproportionate stratified sampling because for all the 14 urban wards (which were the strata- see Table 2.2), on average 13-14 households were interviewed irrespective of population figures in each ward. The goal was to have each ward represented by a minimum number of 10 households

Table 2.2. Overview of selected districts and wards in Mwanza with number of inhabitants, households and households interviewed per ward.

Districts	Wards	Pop. 2002 census	Number of households	No. interviewed
	Pasiansi	25,310	5,410	13
	Butimba	31,109	6,287	12
	Nyakato	82,381	17,410	15
Ilemela district	Ilemela	23,952	4,922	12
	Mkuyuni	13,343	3,416	10
	Nyamanoro	42,731	9,647	12
	Igogo	28,570	7,289	14
	Pamba	23,546	5,130	16
	Nyamagana	5,851	1,236	21
	Mirongo	5,332	1,109	14
	Mbugani	37,522	9,111	15
	Isamilo	17,916	4,096	15
	Kirumba	21,642	4,989	21
Nyamagana district	Kitangiri	14,282	3,115	10
	Total			200

Source: Field work 2007-2009.

and to arrive at a maximum total sample of 200 households. Using the population census data and with the help of the ward leaders and the public health officers posted to each ward, households were randomly picked from a list and questionnaires administered (Table 2.2).

Kisumu

In Kisumu, SWM is patterned much more along income levels. Previous empirical work indicated that the council had not officially permitted non-state actors to operate but all SWM activities by these actors went on unofficially. SWM service providers (the non-state actors) defined their clients on the basis of their income. Community self-help groups were common in low income areas while private companies dominated high and middle income areas. Therefore a list of low, middle and high income estates was made, and then a few estates from each of the three categories were randomly picked

Note that just like in Mwanza, Kisumu has civic wards, some of these wards are estates in themselves for instance, Nyalenda is a ward and an estate at the same time, while in other wards, there is more than one estate. There are in total 17 civic wards covering 41 estates.

There are about 41 estates recognised by the council (11 high income, 17 middle income and 13 low income). The council provides waste management services in only 12 of these estates. The study aimed to administer 200 questionnaires just like in Mwanza. About half the number of estates in each income category mentioned earlier was randomly selected. The study ended up administering questionnaires in 6 high income, 9 middle income and 7 low income estates listed in the Table 2.3. About 10 questionnaires were administered in each of the estates selected.

The number of households per estate varies from about 3,200 to about 12,000. Selecting households within the selected estates was done differently depending on the kind of estate

Table 2.3. Overview of selected estates in Kisumu city per income category and number of households interviewed per estate.

High income	Number interviewed	Middle income	Number interviewed	Low income	Number interviewed
Milimani	10	Dunga	10	Railways	10
Tom Mboya	7	Mosque	10	Nyawita	10
Mt.View	8	Upper railways	10	Manyatta	10
Poly View	10	Makasembo	9	Manyatta Arabs	10
Mayfair	10	Arina	8	Nyalenda	11
Migosi	10	Robert Ouko	11	Shauri Moyo	7
		Okore	5	Nyamasaria	4
		Nubian	10		
		Ondiek	10		
Total	55		83		62

Source: Field work 2007-2009.

arrangement. In planned estates like the Railway estates, the houses are numbered and organised in a certain pattern so it was easy to do systematic sampling, selecting every fourth household. In the informal estates like Nyalenda, the houses are not numbered or arranged in any particular order, the researchers were guided by the village names within Nyalenda which were listed and then one household from each village was randomly selected. To avoid covering a village more than once, the study used one research assistant per estate.

The income categorisation used to obtain the sample reflects the general pattern of service provision in the town. The estates were classified according to one of the three distinct income categories. From each group of estates about half was randomly selected. Within each selected estate about 10 households were randomly selected. Therefore, based on the argument of inferential statistics, the data can be considered to reflect the situation of service provision among the different income categories in the town as a whole.

Jinja

In Jinja, there are three divisions (Table 2.4). Within each division there are parishes but solid waste management services have been contracted out per division and the work is given to two contractors. The divisions are:
* Central division;
* Walukuba division;
* Mpumudde division.

One contractor serves both Central and Walukuba divisions while the other contractor serves Mpumudde division. Service is therefore not structured along income levels as in Kisumu neither per ward as is the case in Mwanza.

Waste is collected from skips (collection points) and not directly from the households. Contractors are paid per emptied skip. There are 119 collection points (skips) in Central, 10 in Walukuba and 20 in Mpumudde. Central and Mpumudde Divisions were picked for the study in order to show the differences (if any) in service provided by the two contractors.

Table 2.4. Overview of divisions in Jinja Municipality with number of inhabitants, households, collection points and households interviewed per selected division.

Division	Pop. 2002 census	Number of households	Number of collection points	No. of households interviewed
Central Division	26,698	5,519	119	180
Mpumudde	19,901	4,220	20	38
Walukuba	24,614	6,795	10	
Total	71,213	16,534		218

Source: Field Work 2007-2009.

Frequency of questionnaires administered was higher in Central with about 180 questionnaires randomly distributed and 38 randomly distributed in Mpumudde. These frequencies were more or else in line with the distribution of collection points which are many in Central division. In the end, in total 218 household questionnaires were administered, so the aim to get a total minimum of 200 questionnaires, as for the other two towns, was reached.

With a list of street names (also referred to as roads/avenues/zones which are equivalent of the villages in Kisumu), households (numbering up to 10 in certain streets) were randomly selected from each street.

For all three towns, questions asked included the income status of the household; who the service provider is; whether the household pays for service provision; how much they pay for SWM; whether they consider the service satisfactory; which among water, electricity, security and SWM as services would they pay for first, second, third and fourth (see Annex 1 for the questionnaire). This was done to assess, among other things, service distribution to conclude on accessibility, different service providers to conclude on flexibility and the relationship between households and service providers on legitimacy and social trust. The household data collected were analysed using SPSS-PASWStatistics_17.0 from which frequency distributions were made which added onto the descriptive parts of the study findings. Qualitative data were coded guided by the dimensions of the conceptual framework and this made it easy to categorise and compare remarks from different interviewees.

2.5.3 Internal validity of research

Yin (2003) gives four tests that are commonly used to establish the quality of any empirical social science research. Prominent among the four is the test for internal validity. Internal validity means basically that you really measure what you want to measure. Triangulation is one of the ways used to ensure internal validity of social research (Meijer *et al.*, 2002; Modell, 2005) and it refers to the use of multiple methods and measures of an empirical phenomenon in order to realise a more accurate analysis and explanation of it. It can occur with data, investigators, theories and even methodologies. In the context of data collection, triangulation serves to corroborate the data gathered from different sources (Yin, 2003). This study gathered data from different sources including documents, resource persons in the field, from households and also from observations intended to explain certain issues in the study. This study used various methods of data collection to get more reliable explanations for certain issues. These research methods include open interviews with resource persons, household surveys, stakeholder workshops and a round table discussion and finally direct observations.

Chapter 3.
Municipal authorities and performance of solid waste management tasks

3.1 Introduction

Solid waste management at the municipal level involves a number of crucial tasks which have both technical and social dimensions (Oosterveer and Spaargaren, 2010; Spaargaren *et al.*, 2006) that run along the waste chain from collection to the point of disposal. While it is obvious that SWM infrastructure has a technological dimension, they need to be implemented and managed in order to fulfil their task. At the same time, they have a social dimension as well, because they need to accommodate the (sometimes widely diverging) local cultural practices and to perform in a sustainable manner to prevent negative environmental impacts. In addition, managing these infrastructures effectively also requires some form of coordination between institutions and actors at multiple scales: neighbourhood, city, national-level, and sometimes even the global level (Oosterveer and Spaargaren, 2010).

With numerous reports by scholars and even local media on non-performance of municipalities in SWM, a crucial question is therefore, what is a good solid waste management system at the municipal level? And to partly answer this, this chapter compares and assesses the performance of three urban authorities (Kisumu, Mwanza and Jinja) focusing on their institutional arrangements for the different SWM tasks. These tasks are presented based on the dimensions discussed under the modernised mixtures approach (MMA) on which this study is based (see Spaargaren *et al.*, 2006). These are the technical dimension which represents the technical aspects of SWM tasks along the waste flow reflecting the municipalities' functional capacity in collection, transportation and disposal of solid waste as well as the managerial requirements involved. Then there is the social dimension which has got to do with the actors/organisations bringing out the social aspects of SWM tasks including the relational capacity of municipalities to engage with other public and private actors, using influencing and networking skills to work collaboratively, to encourage public participation and to build social trust. In this regard, the ways in which local authorities arrange their organisation is a vital issue, as these institutional arrangements can provide for, or hinder, their capacities related to both the decisions around, and the delivery of, key services (UNDP, 2009). The MMA aims for optimisation via combining technical and social dimensions, depending on the specific local situations. To allow for assessing the result of this optimisation attempt, the MMA suggest to use the criteria of robustness, flexibility, accessibility and sustainability.

Briefly the research methodology is explained in section 3.1.1, followed by section 3.2 which presents the tasks identified according to theory and their actual performance on the ground. The last portion of this section explains the (lack of adequate) performance of the tasks. Section 3.3 takes the MM discussion a little further by exploring the relevance of municipal autonomy for dealing with SWM tasks by focusing on relations between municipalities and other actors (both state and non-state) at local and supra local levels. Section 3.4 presents the conclusions of this chapter.

3.1.1 Methodology

This study employs the Parallel Multiple Case studies research design already explained in Chapter two. The cases studied are the three urban authorities (Kisumu, Jinja and Mwanza) and the research is parallel because all three cases were studied at the same time and not one after the other. The data that informed this chapter was collected through the application of different methods:

Relevant documents, such as pieces of legislation and policy papers were analysed. In particular, the legislation governing the urban authorities in each of the three countries was studied, followed by the legislation on solid waste management and in particular the specific by-laws on SWM that exist for each council, followed by a review of reports on the profiles of each council.

Semi-structured interviews were conducted. Even though a structured list (see Annex 2) was used of issues to be addressed, the semi-structured interviews allowed interviewees to raise additional and complementary issues some of which will be presented in this chapter as verbatim. Performance of the technical and social tasks by the municipalities were included in the interviews. The interviewees included 12 resource persons from the three municipal offices Others interviewed were non-state SWM service providers who included two contractors in Uganda, 31 informal group representatives in Kisumu and 16 contractors in Mwanza.

Direct observation of the SWM management infrastructure was also helpful in drawing conclusions about their status.

Data analysis is done mostly qualitative. The different SWM tasks form the categories to present the results from the research and they are assessed for robustness, accessibility, flexibility and ecological sustainability.

3.2 SWM tasks for municipal authorities

3.2.1 Tasks identified according to theory

SWM tasks performed at the municipal level have both technical and social dimensions. Looking at the tasks from these two dimensions builds on the many studies on social-technical systems that acknowledge the co-evolution of technical and social developments.

In this section, I present these two dimensions separately as shown in Figure 3.1. First, I look at the technical dimension drawing upon the discussions surrounding centralised and decentralised modes of environmental infrastructure and service provision. Secondly, I present the social dimension from the perspective of the discourse of developmental state vs. network governance in infrastructure and service provision.

Technical dimension

The technical dimensions covers tasks regarding planning, implementation and maintenance of collection and transportation systems, waste recovery and final disposal. Thereby one has to take into consideration the design and selection of facilities and equipment with regard to their operational characteristics, their performance and their maintenance requirements. The need

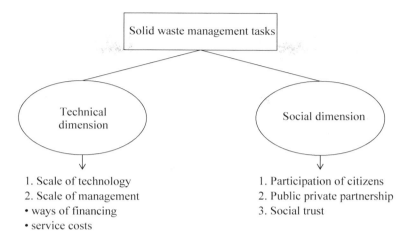

Figure 3.1. Dimensions of SWM tasks.

for repair and the availability of spare parts should not be forgotten either (see Schubeler, 1996). These tasks along the waste chain are covered under the following heading:

1. *Scale of technology.* The scale of technology refers to the scale of infrastructure and service provision which ranges from large-scale which is typical for centralised systems to the small-scale infrastructure and service provision which are usually decentralised. The scale of technology can be clearly pointed out under waste collection, transportation and disposal.

2. *Scale of management.* The scale of management refers to the way in which system operation, management and maintenance tasks are handled and this usually ranges from centralised to decentralised management (Hegger, 2007). Included here are the ways in which SWM is financed and how service costs are collected because the argument has always been that local governments are mainly reliant for funding on central governments and donor transfers that are often conditional, unpredictable and not sustained (Okot-Okumu and Oosterveer, 2010; Rotich *et al.*, 2006; UNDP, 2010). Local governments are generally seen to be weak in revenue mobilisation.

From these two headings, it is clear that we have two extremes and which are centralised technical and management scales on the one hand and decentralised ones on the other. The choice between 'centralised' and 'decentralised' options is certainly a crucial issue and lies at the core of many debates around environmental performance of urban infrastructure and service provision (Tjallingii 1996 in Hegger, 2007). The centralised mode as discussed in Hegger (2007) is seen to have a certain degree of security because the very crucial tasks are in the hands of a few people, in this case the municipal authorities, receiving directives and resources from the central governments; they are also considered robust because of minimal interference from the larger public while they also allow for economies of scale. This centralised and hierarchical organisation is typical of large technical systems which are characterised by the presence of a dominant perspective on the role of technology and how technology should be implemented (Guy *et al.*, 2001; Hegger, 2007). Yet, in the case of East Africa there is reported to exist a considerable mismatch between this approach and the actual situation, because large centralised technological systems make strong assumptions about the presence of homogeneity in housing stock, density, degree of urbanisation, accessibility, related infrastructure (such as street paving and drainage), and the like (Spaargaren *et al.*, 2006). On the other end, there are decentralised initiatives like small scale neighbourhood waste collection by community groups considered flexible and able to reach the poor and unplanned neighbourhoods. Oosterveer and Spaargaren (2010) report that these decentralised systems have been noted to be more robust, cheaper and better able to deal effectively with the existing environmental challenges. However these technologies offer solutions for individual households but they do not solve the massive challenges of addressing the SWM problems of large cities in developing countries because it is challenging to up-scale such technologies. Large cities in developing regions such as East Africa are therefore faced with the dilemma of which path to choose for improving solid waste infrastructures as both large-scale centralised systems and small-scale decentralised systems each show serious weaknesses. The modernised mixtures approach on which this study is based, argues for getting the best mix out of both the centralised and the decentralised systems.

Social dimension

The social dimension of municipal SMW allows the study to focus on tasks such as capturing the participation of citizens, the possibilities of public-private partnerships and the presence of social trust. The relational capacity of municipalities to engage with other public and private actors, using their influencing and networking skills to work collaboratively is brought to the fore as well (UNDP, 2010). To be effective in this domain, knowledge and expertise are to be coordinated, rather than controlled by a single source of authority. When working in collaboration with citizen groups, a combination of trust and the ability to monitor and exercise some control are needed. However, entering into partnerships with non-state actors in the delivery of core municipal functions and services requires an immense cultural shift for most local governments and a willingness to give up some of their control over material and non-material resources.

An interesting area of debate has therefore to do with the participation of citizens and the involvement of other non-state actors in service provision because this relates to the debate on the neo-developmental state versus network governance. As presented in Chapter 2, proponents of the neo-developmental state claim that it is only through active government interventions that access of the poor to environmental infrastructures can be secured, as their economic and political power is too limited to realise this in another way. While according to the network governance view, (urban) governmental authorities are no longer pivotal in securing urban environmental infrastructures and services, because they have become one among the various societal actors that contribute. Network governance arrangements are seen to create room for effective local communities' participation in developing their own preferred approach while harmonising this aim with other stakeholders and other levels of governance.

Along the same line of non-state actors, a practical livelihood aspect that nuances the discussion on the social dimension of municipal SWM is the issue of social trust. Social trust is difficult to define precisely, but it has been encapsulated as an ongoing motivation or impetus for social relations that form the basis for interaction (Jordan, 2005). Social trust can entail perceived honesty, objectivity, consistency, competence, and fairness, all of which foster relationships between individuals/parties that must be maintained by the sustained fulfilment of these standards. In the field of municipal SWM infrastructure and service provision, this can be reflected by sustained or consistent service provision which is accessible to all, irrespective of their income levels. Of particular importance here is to distinguish between 'active' and 'passive' trust. As Hegger (2007) explains, active trust (for example in public figures, institutions and certain others) can no longer be taken for granted, as it used to be (i.e. passive trust), but has to be negotiated. Trust needs to be continuously reconfirmed, and may be withdrawn if it turns out to have been given unjustified. In those cases, actors may direct their trust to others (e.g. other persons, or other institutions). It cannot be determined a priori whether a public or privatised modes of provision is most conducive to trust building, as what matters most is whether mutual expectations are met and people know what to expect from each other. Poor levels of service, interruptions, low coverage levels, and other problems can undermine the quality of life in municipalities and erode public trust in the local government. On these grounds, social trust is included in this study and it is presented in the empirical section under the question whether the service providers, and particularly the municipalities, have invested in social trust. Though not measured directly, this social trust is reflected in the level of satisfaction about services and the level of public participation.

On the basis of this theoretical understanding, the next section presents the empirical evidence that will allow the study to conclude on the feasible options for municipal SWM as concerns the performance of their tasks under both the technical and social dimension

According to the MM approach it is essential to explore the municipal autonomy to ensure that the local solutions identified are considered potential building blocks for larger responses and dealt with in the framework of the complete waste chain, from collection to final disposal, and at multiple levels from local up to the regional and international levels. This allows the study to further assess the flexibility of the MM approach.

3.2.2 Tasks performed

Technical dimension: scale of technology

Collection

The collection of municipal solid waste varies between the three towns. In Kisumu there is a centralised mode of collection by the municipal authority which is completely detached from the informal/unofficial arrangement by the private firms and community groups. During the field work, the Deputy Director of Environment informed the study that the informal service providers collect 10-15% of the waste generated. This is a significant amount given that the council which is charged with the responsibility for waste management is only able to collect 25-30%. The frequency of collection varies between the council and the private collectors. The private collectors have their own arrangement with the clients ranging from collecting daily to only once a week. The council collects daily from the central business district (CBD) and the main municipal market and once a week from certain residential areas. For instance in the Nubian Estate they collect on

Saturdays, in Tom Mboya estate every Tuesday, in certain areas of Manyatta and Milimani, every Wednesday while in the Russian quarters, they collect every Friday. Both the *1954 Kisumu By-laws and the 2005 draft* require that all business, residential, entertainment and recreational premises have a waste receptacle for storage of the wastes generated before collection and that all landlords must provide waste receptacles and ensure that their tenants do not dispose of their waste in the open. These instructions are not put in practice on the ground. While the CBD and the markets have litter bins and collection chambers respectively, the private companies supply plastic bags and some of the community groups use sacks and buckets to do the collection which are easy to transfer to their collection point or disposal site using small vehicles. Overall, the percentage of waste collected remains very small, and this could be attributed to a number of reasons. One of the main reasons is the detached, informal arrangement implemented by the private companies and community self-help groups. Private companies only cover the high and middle income areas where they receive payments for the service provided, but the council's services are also concentrated in the CBD leaving the low income areas to the community self-help groups who only offer limited services. Another reason for this weak performance are the limited resources available to the council and this is explained further in subsequent sections in this chapter.

In Jinja, even after a process of privatisation has been implemented, the collection system still follows a centralised mode. A contractor collects waste from households and commercial enterprises by means of skips using council vehicles and monitored by a council officer. The contractors are able to collect between 40 and 60% of the waste generated, a figure slightly higher compared to Kisumu which can be explained by the formal contracting arrangement on waste collection as opposed to the informal arrangement in Kisumu which is not linked to the municipal authority. Therefore the nature of linkages between municipalities and the formal/informal status of service providers influences whether the system is of a robust nature. Amongst the three divisions in Jinja, there are 119 collection points in Central, 20 in Mpumudde and 10 in Walukuba Division. This means that service provision is concentrated in the Central Division and that households in the other two divisions have limited access of the service so they may end up practicing backyard dumping and composting. The frequency of emptying skips depends, among other considerations, on how fast the skips fill up but in general the collection is done daily, meaning at any one day, there is a skip that is collected and emptied. The contractor serving Central and Walukuba claimed that the company collects and empties on average 18 skips per day while the other contactor serving Mpumudde reported to emptying on average 7-8 skips per day. This number of skips emptied is small compared to the number of collection points available and this can be explained by the limited resources available in the council since according to the arrangement the contractors use council vehicles for collection.

In Mwanza, centralised and decentralised modes of collection are combined. Waste collection has been privatised which is a form of external decentralisation but for every zone to which a contractor is assigned, there is also a council officer who supervises their performance. Since privatisation, the percentage of waste collected has gone up from 28%[5] in 2003 to 88% in 2007 (field

---

[5] Information obtained through an interview with the solid waste manager and corroborated by SWM data of Fredrick Salukele (PhD Wageningen University). See also ILO, 2006 which gives collection rates of 28% in 2003, 47% in 2004 and 80% in 2005.

data, 2007-2009). The frequency of collection varies between the different wards. In the wards of the CBD, waste is collected from transfer stations daily and sometimes even twice a day because in addition to households business activities are going on there and they generate a lot of waste. In the other wards, waste is collected at a maximum of three times per week. While the *Refuse collection and Disposal by-law Article 4 and 5* requires that every landlord/tenant has a standard dustbin for storage of waste before disposal, it was observed that only the private companies give waste containers to their clients. In the other wards people use whatever containers available to them to store their waste. A ward has at least two transfer stations which are either in a form of an open ground collection area or a skip bucket. Mwanza has a total of 30 skip buckets. The high percentage of waste collected in Mwanza compared to the other two towns can be explained by the formal incorporation of CBOs and private companies in the arrangement. CBOs serve different zones according to their capacities, while the private companies have been allocated the CBD area because they have more resources compared to CBOs. Private companies are able to collect and transport the waste to the dumpsite while CBOs are relying on municipal vehicles. These non-state actors bring their resources with them to complement the little that municipalities dispose of and in turn ensure the continuity of service provision and a system that is accessible to all.

Transportation

The municipal transport system for all three councils can be categorised as large technical infrastructure when compared to the small pick-ups and wheelbarrows used by non-state actors. The transport system is characterised by the utilisation of imported vehicles provided by international donors or otherwise sourced from areas beyond the local scale and this has direct impacts on the efficiency of operating and maintaining these vehicles (Table 3.1).

Table 3.1. SWM vehicles in the three councils.

	Type of vehicles	No. vehicles in use	No. vehicles broken
Kisumu	Bulk compactor	1	0
	Truck lorries	2	0
	Tractor	1	0
Jinja	Truck (Benz)	1	1
	Tippers (Tatas)	2	0
	Tractor scooper	1	0
	Bulldozer	1	0
Mwanza	Skip loader	2	0
	Side loader	4	0
	Small vehicles for monitoring	2	0
	Excavator	0	1

Source: compiled by author based on data provided by the council between 2007 and 2009.

For Kisumu (as indicated in Table 3.1), there were no cases of broken vehicles, however the deputy director informed the study that the bulk compactor is very old, donated in the 1980's, and unable to be put to work continually especially with the increasing weight of (particularly organic) waste. Repairs and spare parts of the vehicle are also a problem as it is difficult to source these locally. The loading height of the two open trucks is very high and since they do not compact, they carry less than two tones with much dirt flying out. The deputy director further explained that in some cases the vehicles have been used inappropriately:

> *We have had instances where these municipal branded vehicles for waste collection have been seen transporting building materials or used to shift peoples' belongings when moving house and therefore creating some kind of competition with their use for waste transportation.*

These issues reduce the efficiency of the vehicles and result ultimately in the small percentage of waste collected. The informal waste collectors, particularly the private companies, use pick-ups which is contrary to the provisions[6] in the law because they are not ecologically sustainable as dirt easily flies out, but these small vehicles can easily access the unplanned neighbourhoods with their small access roads. Most of the self-help groups use handcarts and wheelbarrows to transport waste from the households to the transfer points while those operating near the dumpsite, take the waste directly from the households to this dumpsite.

In Jinja, the contractors hire and use council vehicles with council drivers – a mixed arrangement but still highly centralised. Only three vehicles were in use at the time of the field study, two by the contractor serving Central and Walukuba and the other one by the contractor serving Mpumudde. Apart from evidence for a collection rate of just about 50%, the contractors complained about the inadequacies of the vehicles. They are paid per number of skip emptied, yet the Benz truck can only take the 3.5 ton skips and the *Tatas* take on the smaller skips which are 3 tons. The Benz truck is broken down and may remain in that position for a long time because according to the contractual arrangement the contractors are expected to take care of the major and minor repairs needed for the vehicles as well as paying for the fuel. Yet from the field work, it has become evident that these contractors are only small companies who were previously CBOs and are unlikely to have adequate resources to pay for major repairs unless the council intervenes. The study was informed that the Benz truck was sourced from Germany and the other vehicles from India, so if the spare parts cannot be sourced locally, broken vehicles remain grounded for a long time.

In Mwanza, the council operates a refuse collection fleet made up from the trucks shown in Table 3.1. The skip loaders carry approximately a load of 7 tons per trip and the side loaders have

---

[6] According to environmental Management and coordination (Waste Management) Regulation 2006, a person granted a license to transport waste shall ensure that:
the collection and transportation of such waste is conducted in such a manner that will not cause scattering, escaping and/or flowing out of waste;
the vehicles and equipment for the transportation of waste are in such a state that shall not cause the scattering of, escaping of, or flowing out of the waste or emitting of noxious smells from waste.
The Municipal council of Kisumu Environmental by-law 2005 requires that an operator shall not cause any waste to spill in unwanted areas during the course of his operation.

a carrying capacity of 7 tons as well but they manage to carry only approximately 4.5 tons per trip due to the density of waste. These vehicles only manage to collect 30% of all waste (Mwanza City Profile). The study was informed that the council planned to add 2 more skip loaders and 3 more side loaders. The current arrangement in Mwanza, however benefits from the input of private companies who transport the waste they collect using their own vehicles and since these private companies serve wards in the CBD and market, they are able to collect more than the council. This flexible and mixed arrangement in transportation has allowed for 88% of the waste collected to be transported to the dumpsite which is very high compared to the other two towns.

Disposal

All three towns use centralised open dumping grounds which are operated by the municipal authorities.

The dumping site in Kisumu is about 3 kms from the CBD. This dumpsite it is not fenced and located very close to a major supermarket outlet, a major highway (Nairobi-Kisumu Highway) and some residential housing units. Although most of these infrastructural developments came up when the dumping site was already in existence, it is nevertheless ecologically unsustainable because of the environmental impacts associated with uncontrolled open dumping, including leachates, odour, contacts with unhealthy hazards among others. The Deputy Director of Environment informed the study that the dumpsite has been in use for the last 40 years and that the council has been practicing open burning of waste at the site to reduce the volume of the waste and this raises questions as to oblivious nature of the council towards smoke emanating from this practice of burning. There are two officers from the council manning the site and they collect a fee for every private vehicle coming in to dump waste. They charge Kshs100 per entry (USD 1.42).[7] Scavenging is however not controlled or restricted and this poses other health concerns. The Deputy Director explained that the council is in the process of relocating the dumping grounds and by the year 2008, there were five potential sites that had been identified awaiting Environmental Impact Assessment.

In Jinja, the dumping site (Masese) is about 6 km from the CBD. It is an open site where entry is restricted by council officers. Waste other than that brought by council vehicles is charged for disposal. For a 2 ton vehicle Ugshs 5,000 (or USD 3.7) is charged, for a 4 ton vehicle Ugshs 10,000 (or USD 7) and Ugshs 30,000 (or USD 21.4) for each vehicle above 4 ton. This arrangement ensures robustness of manning the disposal grounds compared to the standard 'small' fee charged in Kisumu. Unlike the dumpsite in Kisumu, Masese is a bit 'hidden' from the busy residential areas and scavenging is not common. It remains however, an ecologically unsustainable practice of disposing waste. At the beginning of 2009, the study learnt about and observed the initial phase of a composting plant that would use the windrow technology at the dumping site. This is an initiative of World bank and the National Environmental Management Authority (NEMA). The Health Inspector overseeing SWM in Central Division informed the researcher that once the construction work will be complete, the plant will be run by the municipality. This is an ecologically sustainable venture and the Director of Environment explained that NEMA expects the compost

---

[7] USD 1 is considered to be equal to 70KsHs; 1Kshs is calculated for 20Tshs and 20 Ugshs.

project to be self-sustaining and that at one point the council would be able to generate a revenue from the sale of compost.

In Mwanza, the dumpsite (Buhongwa) is about 18 kms away from the CBD, a distance that can be considered relatively far for small-scale private entities with limited resources. Though it is an open ground, it is partially fenced and has a small office building at the entrance for the two dumpsite attendants who are employees of the council. At the entrance, vehicles pay the fee shown in Table 3.2. The Mwanza arrangement with a stationary office is exemplary compared to the other two because it signifies the management attention given to waste disposal. Just like in the other two towns, however, the open dumping is not ecologically sustainable but at the time of the field work, the study observed some form of small-scale plastic waste collection at the dumpsite for recycling into reusable products. This was however not a council initiative but done by individuals. The study learnt that the council recognised this initiative but not on paper – it is a kind of 'silent agreement'. This is an initiative which if encouraged by the council, with proper official working conditions could reduce the amounts of waste ending up at the disposal site.

Table 3.2. Charges for waste disposal for different groups in Mwanza.

Industries	Institutions	Organisations	Individuals	Others
100,000 Tshs (USD 71) per trip	30,000 Tshs (USD 21) per trip	50,000 Tshs (USD 36) per trip	10,000 Tshs (USD 7) per trip	40,000 Tshs (USD 29) per trip

Source: Field work in Mwanza.

Technical dimension: scale of management

Managerial organisation

In all the three councils solid waste management is run by particular departments. This is a form of administrative decentralisation to lower units of management but the three different departments depict different modes of addressing solid waste management which in turn reflects on how robust these systems are.

In Kisumu SWM was under the docket of the Public Health Department until the year 2004 when it was moved to the Department of Environment because the classical approach to handling waste has always been considered the responsibility of public health. It is now managed by the Environmental Planning and Management Division though the functions spill over to other divisions in the department and even to other departments in the council stemming from the inevitable fact that environmental issues transgress the domain of one department. Locating solid waste management in the department of environment however, has allowed SWM issues in Kisumu to receive more attention than when it is tucked into a sub-department of another department. Such an arrangement ideally should allow resource allocation for waste management to receive priority just like other services at the council and in turn ensure a robust SWM system.

As concerns the staff in Kisumu, SWM is overseen by a director and from the field interview we found that the director has qualifications (Tertiary Qualification) in environmental matters in accordance to section 125 of the LG Act. This is important in ensuring a robust department. This is concluded with an understanding that such an officer will be able to inject the right skills and experiences to keep the SWM system operating efficiently. Next, there are a deputy director and heads of different sections. The bulk of the staff exists of casual workers but even then (at the time of field work),the waste management section had a deficit of 34 staff members (see Table 3.3) who would be required to provide adequate services. An inquiry as to that finding revealed that the absent officers were either ill or deceased and yet have to be replaced. The hiring and appointment of low cadre personnel follows a committee system. There is the technical committee, then the staff committee whose minutes are read to the environmental committee. The finance committee then has to approve and then the full council meeting has a final say. This procedure can actually be done within a month, however in practice the process faces delays from vested interests and political interference from those in charge, while the municipal authority has some autonomy in effecting the low cadre appointments. This inadequacy in available staff impacts directly on the services provided and the robustness of the municipal system and contributes to explaining the low percentage of waste collected in Kisumu by the council.

In Jinja, solid waste is managed by the department of Public Health found at the Local Council IV. The department is headed by a Medical Officer of Health and serves both the curative and cleansing tasks of public health provision. A health inspector oversees waste management in every division of the council. The health inspector also monitors the operations from the contractors in their respective division, an indication of some form of public-private partnership (PPP). SWM in Jinja however, takes a subordinate position as compared to medical/health issues. During the field study, as one enters the department, a queue of patients waiting to be attended to is the first sight and then one has to inquire where the office handling waste is located. So while environmental concerns are just as important as medical ones, waste management remains of little priority in the department and this affects its institutional sustainability. Okot-Okumu and Oosterveer (2010) record that in the financial year 2006/2007, less than 10% of the total municipal budget was used for solid waste management and sanitation by local governments in Uganda.

As concerns staff, there are two health inspectors in charge of SWM in each of the three divisions. The field interviews revealed that they have tertiary (diplomas) qualifications in environmental health and the contractors who provide SWM services went through training

Table 3.3. Summary of staff positions (cleansing and management section) in Kisumu.

Category	Present level	Optimal level	Deficit
Street cleaning & management	23	46	23
Refuse collection & disposal	12	21	9
General supervision	2	4	2
Totals	37	71	34

Source: Kisumu City Council Offices.

on SWM (by ILO) as part of an induction process in conjunction with the council. There are a few sweepers employed by the council but one of the contractors informed the study that there is a plan to have the sweepers employed under the contractors' arrangement to improve on the coordination of cleaning up areas and collecting waste. Currently sweeping is done by the council and collection of the swept waste is done by the contractors. While recruitment and employment has been decentralised to the local level and done by District Service Commission unlike in Kisumu, the number of staff is not as high because service has been contracted out to private enterprises. Before privatisation in 2004, the council had 125 members of staff working on SWM most of whom were retrenched. Privatisation could be seen to have been effective in reducing the financial load of the council with regard to paying workers. And in reiterating this, on being asked whether this is a better arrangement, one of the health inspector said:

> *It may be seen to have reduced the cost incurred by the municipality in paying these workers because I am aware what the workers are paid by the contractors is much less than what the council used to pay workers for the same job.*

In Mwanza, the Health and Sanitation department is responsible for solid waste management under one of its three sections, the cleanliness section. This section is headed by a solid waste manager. The waste manager oversees the public health inspectors posted to each ward. The public health inspectors work in collaboration with the contractors assigned to the different wards. From the field visits, like is the case in Jinja, it is evident that medical/health issues receive more attention compared to SWM. On arrival at the council offices, there is a big medical clinic and offices handling medical issues at the front yet SWM office is located at the back and a visitor would need to ask to be shown where it is. This in turn also depicts the resources allocated to SWM. The arrangement however also shows flexibility in the mix of public and private actors where the public officers monitor the private initiatives. This is necessary to allow SWM to gain from the benefits that each group brings on board.

In terms of staff, there is at least one public health inspectors (PHI) from the council for each ward monitoring the activities of the CBOs and the two private companies who have been awarded contracts. The solid waste manager informs that the PHIs have academic qualifications and experience before one is appointed, as is required. They are therefore knowledgeable on waste management issues. They report to the solid waste manager who is in charge of solid waste in the city as a whole. The study was informed that there are 44 PHIs in total. Their employment or recruitment is done with permission from the central government through the Prime Minister's office meaning that the council is not in a position to influence this process. In addition to the PHIs, the council has also nine drivers and two attendants at the dumpsite. In the meantime, any inadequacies in the council's capacity to manage waste on its own are partly addressed by inputs from the CBOs and the private companies.

Ways of financing

When looking at the financing of the three municipal councils in general, Kisumu's funds received from the central government makes up approximately 24% of the overall council's revenue. On the other hand, Jinja and Mwanza receive more than 50% of their council funding from the central government. This gives an indication of the dependency of municipal authorities' budgets from fund received from the central governments.

For Kisumu council, SWM receives its funds from the conservancy fee of USD 0.67 charged per household per month through the water bill, but this includes only those households that have metered water connection (that is connected to the central grid). There are also other funds originating from the dumping fee charged at USD 1.4 per load of a pick-up truck. Businesses, particularly the markets and other commercial areas, are charged for SWM through their business license. The other percentage of funds for SWM comes from what is transferred from the national government to the local authority - the Local Authority Transfer Fund (LATF). All these contribution combined are however, not sufficient to adequately run the SWM system for the municipality as a whole. In the financial year 2006/07, the annual income from solid waste management was USD 70,000 against an annual expenditure of about USD 420,000. To address this problem, the study was informed by the Deputy Director of Environment of an ongoing Kisumu Integrated Solid Waste Management Programme funded by SIDA. Also involved are UN-Habitat, ILO, Lake Victoria Local Authorities Cooperation (a regional organisation), Practical Action (an international development charity organisation) and Umande Trust (a Kenyan rights-based organisation). These organisations all handle different components of the programme and through them funds are channelled to ensure transparency and accountability.

For Jinja, the interviews revealed that the municipal authority has a budget for SWM which is mostly funded by the central government. There are also fees collected from business premises (Ugshs 20,000 - USD 14 per business) and a dumping fee at the disposal site but all these go to the central reserve at the council where they tend to be absorbed by overall council expenditures. The Director of Environment informed that the council actually has relative autonomy in putting up its own budget as provided for in the Local Government Act 1997. Yet it is evident that because most of the funds come from the central government and they are conditional, it is unlikely that adequate resources would be set aside for waste management compared to other 'important' municipal services. For the year 2008, the budgeted expenditure for waste management was Ugshs 69,600,000 (USD 49,714) but the actual expenditure came to Ugshs 120,000,000 (USD 85,714) reflecting a deficit of about Ugshs 50,000,000 (USD 35,714) on the budget. Though the council has authority to raise revenues and initiate development projects as provided for in the Local Government Act (1997), the council has not fully exploited this opportunity. Households do not pay for waste collection leaving the council to pay the contractors from the Property Tax which is not very reliable. The contractor serving the central Division therefore complained about this form of payment:

> *A lump sum paid at the end of the month is better than the bits of money we are paid from the Property Tax, because at the end of the day, I also have to pay 18% VAT and in case I default, I pay withholding tax of 6%. When I remove my expenses and the total tax, then I am left with very little money.*

In Mwanza, the study learned that the council also gets most of its SWM funds from the central government. There are additional funds from the fee charged at the dumping site and also from the fee charged to CBOs to have their waste transferred to the dumping ground but like in the other two councils, these funds end up at the central reserve in the council. The solid waste manager informed the study that SWM is not properly defined in the council's overall budget but that their total expenditures for the year 2007 went up to Tshs 210,900,000 (USD 150,643). This figure like the one in Jinja is small compared to the expenditures in Kisumu and this could be linked to the fact that the costs incurred in Jinja and Mwanza are shared by the council and the private sector contracted to provide SWM services. Overall however, Mwanza, like the other two councils, has not exploited all the avenues for raising its revenues given that it has the mandate to do so through the Local Government Finance Act of 1982. Its own sources for SWM revenue remain the dumping fee and the fee for transporting waste collected by CBOs to the dumpsite.

Service costs

In Kisumu, the cost of SWM provision charged at household level varies from place to place.

The council has been charging Kshs 40 (USD 0.6) per household per month but this is only for those households that are connected to the central grid-metered water connection. Charges for collection by the private sector range between Kshs 150-500 (USD 2.1-7.1) per household per month, a figure which is way beyond the capacity of the urban poor[8]. CBOs however charge as little as Kshs 40 (USD 0.6) per household, per month. Although informal, this flexibility in pricing ensures a wide coverage of SWM services. What is also interesting to note is that Section 148 of the LG Act gives the council the power to impose fees or charges which shall be regulated by a by-law or if not regulated by by-law, may be imposed by resolution of the council with the consent of the Minister of Local Government. The input from the public in deciding on fees is obviously missing in this legal provision and in the event that the informal waste providers are made formal, then this will impact on the flexibility of pricing that both the public and the informal waste providers enjoy at the moment. The low amount charged by the CBOs could be because they do not pay tax and may not be renting premises but as to whether this is sustainable in terms of them recovering costs is another issue. It also follows that the low charges reflect the quality of service given compared to that provided by the private companies who from the field work serve different clientele.

In Jinja, households do not pay for waste collection. Field interviews revealed that attempts to introduce a fee of Ugshs 2,000 (USD 1.4) were unsuccessful because households are of the understanding that SWM should be provided by the council and for free. This impacts negatively on ensuring that the SWM system remains sustainable. This turn of events is taking place despite the fact that Section 80 of the LG Act, provides for the municipality the authority to levy, charge and collect fees and taxes. At the time of the field work, the officers interviewed were optimistic that the user fees would ultimately be enforced as these are part of the provisions in the Jinja SWM draft bylaw 2005, yet to be passed. For a sustainable system that will allow full cost recovery from service provision, there is need to have the households pay for the services provided.

---

[8] According to the 1999 poverty rate for rural locations and urban sub-locations (CBS, 2003), and the 1999 poverty rate for constituencies (CBS, 2005), the urban poverty line for Kenya is given as Kshs 2,648 per month (about USD 1.26 per day). This means that 48% of Kisumu's population live below the poverty line.

In Mwanza all households pay a standard fee of USD 0.3 per month per household. This fee, the study learnt, was agreed upon in consultation with the households and has served to ensure an almost uniform coverage of SWM service. The LG Act of 1982 in Section 66 gives Mwanza municipality the power to charge fees for any service or facility provided by it or for any license or permit issued by the authority under the Act. Through its by-laws, Mwanza is to constitute fees for waste collection in consultation with the public. This is a notable contrast to the legal provisions governing Kisumu. When compared to the contributions in the municipalities of Dar-es Salaam, the capital city of Tanzania, as provided in their by-laws, of Tshs 500 (USD 0.4) per household per month for low income, Tshs 1000 (USD 0.7) for middle incomes and Tshs 2000 (USD 1.4) for high incomes, the standard uniform amount of Tshs 400 (USD 0.3) charged in Mwanza, can be gradually increased. This rise will also be commensurate with the changing economic times so that the SWM system remains robust.

Social dimension

Participation of citizens

In Kisumu, there was no evidence of active citizen involvement in solid waste management within the municipal structures. There are no legal or administrative structures for community participation particularly at the ward level to influence decision making within the municipal authority. What was evident however, is that citizens are taking it upon themselves to provide SWM services. They have formed youth groups, women groups and private companies to provide SWM services (discussed more in detail in Chapter 5). The law only guarantees political participation through civic elections.[9] There has been marked attempts at using the Local Authority Service Delivery Plan (LASDAP) process as a requirement before disbursement from the LATF. LASDAP is a local level participatory planning and action tool that requires local authorities to hold consultative meetings with various stakeholders and citizens to agree on the broad allocation of the budget. There are questions however, related to the composition of the participants to these sessions, to the extent to which they are representative for the wider public. As part of LASDAP, every local authority is to have a LA budget day but experience from past sessions has shown that such attempts are marred by political intrigues. In the year 2009, the Kisumu Budget day was marked by a heightened tension and tight security as councillors opposed to the sitting mayor boycotted the event. Police had to face off youths who were hired to disrupt the functioning of the session.[10] Its thus unlikely that such a well-intended function could achieve meaningful public participation in practice. For LADSDAP to have an effective contribution towards effecting public participation, it should be institutionalised at the municipal level as opposed to just being a condition before accessing LATF

In Jinja, there are lower level offices within the municipal structure that provide avenues for public participation. Apart from these lower local level administrative offices (Level II which is

---

[9] Once elected, the government remains the key decision maker on local developmental matters. It is assumed that the councilors effectively represent citizens (UN-Habitat, 2002).

[10] Kenya Broadcasting Corporation. Local Authorities Present Budgets. Rose Kamau, 25th June, 2009.

the parish council and Level I which is the village level), the health inspector in Central Division informed the study that they have had cases where the public have complained about poor waste management through the local media and this triggered a response from the council. The chairperson of the village level (LC I) can also report to the health inspectors at the municipal division level (LC III) who deal directly with the contractors in the event that there is a filled up skip that has remained unattended for a long time. This arrangement involving the lower level administrative offices has allowed the public to access municipal authority and be part of waste management to a certain degree and in turn this ensures continuity of service provision to those areas that are already receiving services.

In Mwanza, just like in Jinja, there are also lower level administrative offices in the municipal structure. These are the *'Mitaa'* or neighbourhood level which consists of a number of households and which reports to the Ward Development Committees. When any scheme or program for the development of the ward has been approved by the council chief executive or by the village councils concerned, the Ward Development Committee is required to inform all persons within the ward area about the scheme or program and the date, time or place upon which the ward residents will report in order to participate in its implementation. From the field work, it was evident that households can and do access the ward leaders and raise complaints in case of a problem because waste collection has even been organised according to the ward structure. The ward leader in turn works with the health inspector from the council and the contractor in the ward. This depicts a mixed arrangement involving the legislative and the executive arm of the council together with the private contractor. In case of any default in paying for waste collection, then the contractor takes this up with the ward leader. One Public Health Inspector explained that the households were indeed involved even in deciding the appropriate location of transfer points.

Public-private partnerships/involvement of non-state actors

In all three councils there is some form of private sector involvement either formal, informal, or both. Apart from this involvement in practice, some provision for it is also made in their respective legislation. There is no national policy on privatisation in the three East African countries but different pieces of legislation allude to it.[11]

In Kisumu, there is evidence from the field work of participation from private companies and community self-help groups (youth groups, women groups) in collecting waste from different neighbourhoods. Their participation is however not formalised as such to warrant calling it a PPP. As will be discussed in chapter 5, the municipal authority is however aware of the existence of these groups and some have actually benefitted from services offered by the council like transporting the waste they collect to dumping grounds or dumping the waste at the disposal site at no cost. While these groups, particularly the community self-help groups, are registered by the ministry

---

[11] In Kenya, the Local Government Act Cap 265 section 143 mandates local authorities to enter into contracts. In Uganda, the Local Government Act of 1997 gives urban authorities autonomy over their financial and planning matters. All urban councils have power therefore to contract out services to the private sector. In Tanzania, Regional Administration Act No. 2 enacted in 1997 provides for the decentralization of municipal services from local authorities to the private as well as the popular sector including individuals, private companies and local communities.

in charge of community development, the registration is purely an administrative exercise as the statutory laws do not provide for it. Registration under this administrative system does not give to the groups any legal personality and neither do the groups acquire corporate identity under the statutory laws. This lack of legal and corporate personality notwithstanding, these community self-help groups together with the private companies as will be shown in chapter 5 have actually scored positively where municipal services are not yet in place. This is particularly the case amongst the low income groups where legal personality and corporate identity in terms of statutory law seem to have relatively little relevance. Yet for the continuity of such an arrangement, it is necessary to link the operations from these community organisations and private companies with the municipal authority. The community self-help groups may need to register as a cooperative society as provided for under the Cooperative Society Act Cap 490 for them to advance the economic interests of their members because as it is, they have no legal authority to compel their clients to pay for services rendered.

In Jinja, there is some form of PPP, because SWM has been privatised and contracts are awarded. At the time of this study, contracts had been awarded to two small private companies. It is actually a real form of partnership because the private companies hire and use council vehicles to collect waste. Besides, the drivers of these vehicles are council employees paid by the council. The council has also employed workers who do the sweeping and make sure waste is put in certain designated areas awaiting collection by the contractors. The council officers also monitor the operations of the contractors on a daily basis. As mentioned above, these contractors are paid by the council from the returns from the Property Tax and during the field interviews with the contractors, there were complaints about delayed payments by the council. Therefore this arrangement, though promoting flexibility by allowing the involvement of the private sector, needs to consider changing the arrangement so that SWM remains robust. The director of Environment actually reiterated this by saying:

> *If the municipal does not create a separate fund for refuse collection or go franchise, the problem with contractors will continue.*

Mwanza has also incorporated the private sector, in particular CBOs and private companies. The arrangement actually displays some form of PPP because the council works hand in hand with the contractors. It transports waste collected by CBOs from the transfer stations to the dumping site though at a fee. Council officers also monitor the operations of the contractors and as mentioned earlier, the council helps to address issues concerning payment defaulters. The Mwanza arrangement reflects more flexibility and potential for remaining robust because the CBOs and the private companies have been awarded contracts in different zones according to their capacity to deliver and households pay the contractors directly for waste collection.

Social trust

Social trust was not directly measured in the three towns but is reflected in the levels of satisfaction which are detailed in the next chapter and the extent of public participation and service coverage discussed above. From the field work, despite the presence of lower level administrative offices in

Jinja and Mwanza, there was no evidence of direct investment in social trust per se by the three municipalities. If the public expresses trust in municipal authorities, it is certainly not because these actively invested in it. While these authorities could be excused because governmental action is bounded by the canons of public law and laws and rules framed under them are often not sufficiently versatile to enable customised responsiveness, this is an area where input from the government could play a great role in improving the relationship with the public and in turn improve SWM. This input could be manifested in different forms including direct investment in public participation and ensuring accessibility to quality and reliable SWM service by all at costs that reflect the local circumstances.

At the time of the field work, in Kisumu, regular service provision by the municipal authority was skewed towards the planned estates and the CBD, which are mainly high and middle income areas. The community self-help groups had to build trust with the low income earners to whom they provide services and one group interviewed in Manyatta explained that they had to offer the service for free for the first six months before they could introduce a user fee.

In Jinja, service is concentrated in the Central Division where the central business district is found, although this does not necessarily signify a concentration according to income levels. The availability of skips where waste is collected is higher in the CBD. There is the 'entitlement culture' and subsequent non-payment by households for waste collection which reflects their social trust towards the municipal authority. Because as earlier mentioned, households believe it is the responsibility of the council to provide services and to provide them free of charge. The principal public health officer in whose docket SWM lies, affirmed this when he said:

> *The public has a low opinion over management of waste thinking the council is responsible; even for dumping rubbish outside the skip.*

The fact that the contractors are only able to provide service to certain areas contributes to painting a picture of low social trust from the public towards the municipal authority.

In Mwanza, there is almost universal service coverage. The city has been divided into three sections:
- section one has Nyamagana and Pamba wards and this is the CBD;
- section two has Mbugani, Isamilo and Mirongo wards and these are the planned areas with more affluent residents;
- section three covers the remaining wards which include planned, unplanned and newly developed areas for the affluent. They are located much further from the CBD compared to areas in section two.

Unlike the situation in Kisumu and Jinja, in Mwanza every ward in the different sections receives services because each ward is served by a particular CBO that has been awarded a contract. The CBOs are made up of groups of local community members who provide services to the wards they live in. This could act to further build the social trust that the public has towards the CBOs and in-turn the municipal authority that contracted them thereby ensuring the building of a robust system.

3.2.3 (Lack of) performance explained

Table 3.4 summarises the results from the performance of SWM tasks presented in this chapter.

Looking at the performance of tasks under the technical dimension within the municipal structure, reveals that all three cases apply a centralised mode of operation that needs to be implemented through under-resourced municipal authorities, particularly concerning the availability of vehicles for collection and transportation. These councils have vehicles sourced from outside their countries whose spare parts cannot be easily found locally. Only in Mwanza the private companies use their own vehicles for collection and transportation of waste. In Kisumu, initiatives by the private sector remain unofficial/informal and in Jinja the skips for collection are unequally distributed. The existing arrangement of collection vehicles and the case of skips in Jinja, as explained in Spaargaren *et al.* (2006), makes assumptions about homogeneity in the housing stock, the density of housing, the degree of urbanisation, the accessibility of vehicles, the existence of related infrastructures, thereby creating a mismatch between resources provided and the actual situation on the ground.

The scale of management in all the three councils depicts a decentralised mode of approach to departments. The managerial organisation in Kisumu however, where SWM is located under the department of environment, serves to give more emphasis to waste management as compared to those in Jinja and Mwanza under the departments of Public Health where medical concerns receive priority. Though this attention is not evident in the amounts of waste collected or the resources available for SWM in Kisumu city council, the interest from major organisations like

Table 3.4. Summary of performance of SWM tasks.

Dimensions	Kisumu	Jinja	Mwanza
Technical dimension			
Scale of technology	Centralised and informal small scale	Centralised	Central and decentralised
Scale of management	Decentralised to department.	Decentralised to department	Decentralised to department
Ways of financing	From central goverment From other sources	Mainly from central government	Mainly from central government
Service costs	Municipal: low cost (fixed) Private: flexible but informal	Non	Low cost (fixed)
Social dimension			
Public participation	Informal initiatives LASDAP not institutionalised	Evident public participation within council structure	Evident public participation within council structure
PPP	Informal PPP	Formal PPP	Formal PPP
Social trust	Evident between public and community self-help groups	'Entitlement culture'	Evident between public and all service providers

SIDA and UN-Habitat in Kisumu's SWM is an indication of the kind of attention SWM in Kisumu is receiving. The inadequate numbers of staff in Kisumu is linked to the fact that the department of environment has to await direction from the top committees charged with appointments. Appointments of low cadre officials are influenced by vested interests but also pegged on the financial resources available. For senior officials though, the ministry of local government has the upper hand.

Under aspects of financing SWM, and cost structure, the three councils depict different performances. For all three councils, funding for SWM for which a high percentage is transferred from the central government to local authorities, is not adequate. As this transfer is also conditional this limits the relative autonomy of the councils in deciding on their immediate needs when using the funds. Where local councils have relative autonomy over their budget like in Jinja, the funds are not adequate enough or not prioritised for SWM. Kisumu benefits from the input from SIDA but for all the three councils, there is need to build on own-source revenues to remain robust and sustainable, for instance, by exploiting waste to energy options or the composting initiative that is coming up in Jinja (see Cointreau, 2005). Finances for SWM should be more prioritised and sourced at the municipal level compared to other municipal services like heath care which could continue relying on the central government funds. This is because local autonomy in SWM allows for more flexibility in addressing the challenges in terms of technology and the actors involved.

The way customers pay for SWM services in Kisumu depicts a certain flexibility in order to capture what different income groups can pay for waste, but again this needs to be regulated so that such services reach a wider group of households. The water meter billing remains restricted to only those who have metered water connection. Jinja's low average performance in waste collection could also be linked to the fact that households do not pay for waste collection and therefore there is need to introduce user charges or collection fees. Mwanza's low standard fee can be progressively increased, as is evident from the case of Dar-es Salaam, so that the SWM system remains robust.

When it comes to the social dimension focusing on participation of citizens and PPP, the three councils record some degree of success in the different aspects. Both Jinja and Mwanza municipal authorities have certainly scored positive on public participation, because the municipal structures have lower level administrative offices that the public can easily access. The LASDAP process in Kisumu may show positive output provided it is institutionalised, because that will translate into the creation of proper structures and offices to guide the selection of participants and the conducting of public sessions.

There is some evidence of a move towards network governance for SWM tasks though it is clear that municipalities retain a strong hold over the operations. It may be necessary to monitor the operations of the non-state actors and also ensure that the poor neighbourhoods receive services as well. There is some form of PPP in these municipalities. PPP in Mwanza and Jinja shows levels of success because here private actors are officially/formally recognised by law and in practice by the municipal authorities. In Mwanza, the contractors are even authorised to collect user fees from the households depicting some form of external decentralisation. In Kisumu, though provided for in the legislation, the private companies and community self-help groups providing service do this informally. Hereby they illustrate the existence of well elaborated, laws and regulations which are not enforced, the so-called 'enforcement gap.' In the end, this analysis shows that while PPP may mean a change in the role of the municipalities, this does not reduce the scope of municipal authority intervention because it requires new forms of regulation and accountability to ensure that private interests accommodate the official state policies (see Bach, 2000).

The three municipalities do not show evidence of direct investment in social trust. The situation in Kisumu depicts low social trust from the public towards the municipal authority because SWM services do not reach most households. Households in Jinja still refer to an 'entitlement culture' and that about 50% of waste generated is not collected, thereby also depicting a low level of social trust in the municipal authority. Even without direct investment, one can conclude that Mwanza municipality enjoys considerable social trust from its residents because not only has collection level gone up to 88% but the contracted CBOs are members of the wards they serve. This as explained by Hegger (2007), acts as a trust builder because information about the functioning, efficiency, levels of achievement, failures and pathologies of groups is then easily available.

3.3 Exploring municipal autonomy under modernised mixtures

To this point, it has been made clear that the tasks of managing SWM by the municipal authority are becoming a 'team sport' that have spilled beyond the borders of government agencies and now engage a far more extensive network of social actors, public as well as private, local as well as supra-local, and, according to Kumar (2004), their participation must be coaxed and coached, not commandeered and controlled.

The concept of autonomy is commonly understood as freedom and capacity to act (Lundquist, 1987). The discussion of municipal autonomy here, derives from the empirical evidence about the involvement of different actors under SWM which has always been a domain of municipal authorities at least for the case of these East African towns. Looking at the relationships between municipalities and these other actors (both state and non-state actors) allows this study to formulate conclusions on the impact these relations have on the capacity and autonomy of municipalities in dealing with SWM tasks.

The relationships between municipalities themselves and between municipalities and non-state actors at the municipal level is obviously key to service provision with a focus on the amount of control, competition, coordination, cooperation and collaboration. These relationships are discussed in detail in Chapters 4 and 5.

When it comes to the relationships between municipalities and agencies beyond local level, it is evident from the empirical work so far that municipalities are supervised by 'superior' levels of government. This supervision is evident in the support given, different interventions, control over budget and control over administrative affairs by national governments. Steytler (2005) explains that the level of supervision inevitably defines the level of autonomy.

The arguments of MMA seek a balance between centralised and decentralised organisation. In the three cases presented under section 3.2.2, central governments display the higher power. From a design point of view, the local governments systems have the requisite legal (and constitutional for the case of Jinja) and institutional frameworks from which to operate and to operate effectively. Though the decentralisation process going on in Kenya, Uganda and Tanzania differs, on a general level, the municipalities within the three countries have been given the mandate to manage waste. They have also been given resources and for the case of Jinja, recruitment of personnel is even done at the local level. However, that over 50% of the financial resources are from the central government and conditional, translates to the fact that, these municipal authorities cannot decide themselves how such funds will be used. And for Kisumu and Mwanza where (senior) personnel are appointed by national bodies, such personnel are likely to have no personal stake in the success of operations at the municipal authorities. The degree of autonomy and capacity of these municipal authorities is thus checked directly from the national level. Though the study at this point is inclined to propose options for more municipal autonomy, it is within the MMA discourse that the allocation of powers and functions to municipalities need not be at the expense of states. Steytler (2005) writes that often strong local government is good for state government. What is required however is clearly defined complimentary roles for each sphere. While local government is crucial in the development and well-being of any country, it cannot be left completely to its own. Supervision is vital but the appropriate balance between supervision and autonomy should be struck.

The relationship between municipalities and regional, east African, actors is discussed in detail in Chapter 6. It presents two regional institutional arrangements and explains the autonomy afforded to municipalities under the different regional arrangements.

Relations between municipalities and the international level are evident in the World bank initiative of composting in Jinja and the input from SIDA towards Kisumu Integrated Solid Waste Management Programme (KISWAMP). Contrary to what Mitchinson (2003) in Oosterveer and Van Vliet (2010) explains that international aid programs seem to move away from project-based interventions (which allows for local government support) toward budget support that shifts decisions back to the centre, these two cases actually display project based interventions. From the field work, upon completion of constructing the compost plant in Jinja, the project will be given to the municipality to run it. This reflects some degree of autonomy for the international actors at the point of infrastructure construction and strengthening the managerial autonomy for the municipality once the construction is complete. Just as is the case at the national level, there is need for balance when it comes to international interventions to allow municipalities some measure of autonomy but also to allow the international actors some autonomy as they play crucial roles not only in the provision of funding, capacity building and technical support, but also in pressing for environmental issues to be at the top of the development agenda. DFID (2005) reiterates this when it says that approaches for donor support should be adopted in a manner that does not undermine the role of the state *(municipality)* but seek to build on where the state can operate (for example in aspects of regulation). Municipal operations can be supported through other organisations as exemplified in the case of SIDA in Kisumu working through ILO, UN-Habitat, Umande Trust, Practical Action and Lake Victoria Region Local Authorities Cooperation, which all play different roles towards effecting the SWM but with UN-Habitat as the lead agency to oversee the program and avoid any conflicts in the performance of the tasks.

3.4 A look into the future

The modernised mixtures approach stands for combining technological and social dimensions, integrating the management of waste flows from collection to final discharge, and linking the different technological and social aspects involved in the provision of environmental infrastructures and services. SWM systems should be seen as socio-technical systems which have become strongly techno-economically, institutionally and socio-culturally embedded.

Improving the performance of local municipalities in SWM tasks therefore calls for a broad and nuanced view of handling different SWM aspects. Municipalities will have to become more strategic in their orientation, be open and flexible to changing circumstances and maximise integrated capacity from both the centralised and decentralised options of service provision. Use should be made of both the large scale and the small scale infrastructure and services as situation permits.

Robust and sustainable institutional arrangements are those that integrate with non-municipal actors while at the same time ensuring that the neo-developmental role of the state is fulfilled effectively. Integrating with non-municipal actors also ensures accessibility of SWM services for most if not all people. Where PPP has been practiced, it has yielded positive results in terms of increased service coverage and increased amounts of waste collected and transported to the disposal site.

Building and maintaining social trust of the wider public would require an assurance of not only accessibility to SWM services but also sustained service provision and their (public) participation particularly in decision making. This translates to the condition that municipal authorities like their private counterparts need to invest in social trust.

Flexibility in the organisation of SWM should be balanced (not too much flexibility) to ensure that all actors have the needed relative autonomy to make decisions and exercise control over infrastructure and service provision. Such flexibility should not be realised at the expense of the quality of the SWM services.

Chapter 4.
Municipal authorities and non-state actors

4.1 Introduction

A study of governance and urban waste must examine not only the formal structures of government but also the informal structures created by society, such as community-based institutions, associations, and organisations, their relationships, and the relationship between the formal and informal structures for collection, transportation, and disposal of waste.

To advance this discussion, this chapter builds on the debate of neo-developmentalism versus network governance which was briefly captured in Chapter 3. Several authors, as well as large sections of the public opinion in Africa, criticise the suggestion of a minimal state and insist on a return to an active involvement of the government as the promoter of development. On the other end, there are scholars who consider the current African state fundamentally incapable of delivering services in a reliable manner and therefore in order to adequately implement and manage urban environmental infrastructures the active involvement of private companies, non-governmental organisations and local communities should be promoted.

In the quest of continuing to answer the question of what a good solid waste management system at the municipal level should be, this chapter compares the involvement of non-state actors at the municipal level in the three towns (Jinja, Mwanza and Kisumu). Based on empirical findings presented in Chapter 3, the study presents the arrangement in the three towns following three different models: a market-like arrangement for Jinja, a community-dominated model for Mwanza and a hierarchical arrangement for Kisumu.

Section 4.2 gives a theoretical introduction to the chapter. Section 4.3 brings out the more detailed aspects of the models of markets and networks, community and networks and hierarchies and networks as revealed in the field work. It is in section 4.4 that the question is posed whether it is most appropriate to opt for markets, communities or hierarchies. The answer to this question is reflected in the conclusions of this chapter on formal and informal collaborations and on the position in the debate on neo-developmental state versus network governance.

4.2 Participation of non-state actors in municipal SWM

Although governments are generally expected to provide environmental services such as sanitation and solid waste collection for their citizens, in most Sub-Saharan African cities and as discussed in Chapter 3, the (municipal) governments are not able to provide these services solely by themselves. Without ignoring the lack of material resources caused by poverty, there are other structural causes as well for this failure and they relate to the way in which the role of the state is understood (Oosterveer, 2009). The withering away of the African state and its diminishing capacity and sovereignty in various fields have created considerable room for non-state actors to move into traditional state tasks (Mol, 2001). These non-state actors start developing their own sets of rules or standards to fill 'institutional voids' where rules to guide behaviour are needed but not provided by the state. These interventions and their associated institutional arrangements have resulted in three different models:

1. *Market dominance.* Here the focus is on the arena of economic actors and this relates to the various mechanisms in which economic actors can or cannot cooperate to resolve common problems without distorting the basic mechanisms of the market. In this model these economic actors dominate the scene and when it comes to SWM, this means that the actual mode of organisation is a matter of preference and choice on the side of the client and of competition amongst the service providers

2. *Community based organisation dominance.* Just as the government, society itself can also be a major implementer (Pierre and Peters, 2000). Governments may use organisations in societies to implement programs for a variety of reasons, not least because when these groups do the implementation, this will save the government money and make the public sector appear less intrusive. Furthermore, implementation of programs through social groups enables governments to utilise the expertise from those groups to make better decisions. The basic question which emerges thereafter is the degree of governability of these societal organisations and/or the degree of autonomy these societal organisations should have.

3. *The hierarchical governance mode.* Sub-national governments (local authorities) enjoy some degree of autonomy from the national governments. These local authorities, however have their own constituted 'powers' – they are conceived as the epitome of collective interest at that local level, governing society by imposing the law and through other forms of regulation. It is a hierarchical system of command and control down to the departments and divisions within these local authorities. Even with the presence of other actors in society like the CBOs and NGOs, there according to this model is no need for much flexibility or diversification or informal exchange. There should be a rather strict division between the public and the private.

For the East African states, there is no case where the state is acting in isolation, neither are there cases where the non-state actors take the full lead in SWM. Important questions arise therefore as to whether to strengthen governmental authorities to take their responsibilities more seriously (following the model of the neo-developmental state) and/or to actively involve private companies, non-governmental organisations and local communities in implementing and managing urban environmental infrastructures (following the network governance approach). Based on arguments for the modernised mixtures approach on which this study is founded, especially in the field of waste, building relationships with non-state actors should get special significance since the past performance of the state in this field is not very good, and citizens tend to use this (lack of proper) performance as an important indicator for the legitimacy and authority of the state and in turn the local authorities.

Jinja, Mwanza and Kisumu[12] display significant variation as to how municipal authorities have shaped their relationships with the non-state actors in the field of SWM. These are discussed in the section 4.3 under the headings: markets and networks, communities and networks and hierarchies and networks. Under each section legitimacy and influence on decision making, relations and alliances as well as the payment systems all of which take on a specific meaning are looked into

---

[12] More background information on these three towns is given in Chapters 1 and 4.

as aspects of governance which in turn influence accessibility, flexibility and sustainability of SWM services. Matovu (2002) asserts this by writing that good governance occurs only if there is legitimacy of authority, public responsiveness, public accountability, and public tolerance of other actors with a public character, information openness, and effectiveness in public management.

4.3 Methodology

This part of the study also employs the Parallel Multiple Case studies research design, already presented in Chapter 3. The cases are the three urban authorities (Kisumu, Jinja and Mwanza) and it is parallel because all the cases were studied at the same time and not one following the other. Different data collection methods were used to get the information presented below.

Interviews were conducted with resource persons, with the aim to get data that would help analyse the legitimacy of the groups involved in SWM, the relations and alliances amongst these groups and the efficiency of payment systems. Those interviewed were:
- heads of the departments managing solid waste in the three councils;
- CBOs involved in waste management in Kisumu (17) and Mwanza (14);
- private companies in SWM in Jinja (2), Kisumu (8) and Mwanza (2);
- recyclers in Kisumu (6);
- 2 SWM associations – one in Mwanza and another in Kisumu.

In Jinja and Mwanza, all the representatives of the SWM providers were interviewed, because they were fewer in number and accessing them was easy as they were all formal contractors. In Kisumu on the other hand, as is elaborated in section 4.3.3, not all groups were interviewed, but a selection on the basis of the availability of the interviewee but the researcher made sure to get several of them in each category of (CBOs, private companies and recyclers). So that out of a total number of 68 groups active in Kisumu, 31 groups were interviewed.

A questionnaire-based survey amongst households was conducted in all of the three urban centres. This method is already described in the methodology section at the end of Chapter 2. These surveys helped to get data on which households received SWM service and from which service provider they received this, which households paid for SWM and how they rated the service provided and also whether they would pay for waste as a priority service compared with others to gauge whether households pay for other services better/more regular and that the payment problem is only an issue to do with waste management.

Observations about the actual SWM management infrastructures were also helpful in drawing conclusions about their status.

SPSS-PASWStatistics_17.0 is used to analyse the quantitative data while coding has been helpful in handling the qualitative data.

4.3.1 Jinja - governance as markets and networks

Governance as markets and networks refers to an arena of economic actors with a reduced political sphere in SWM management. The networks aspect is relevant because of the inclusion of other actors who also take part in SWM issues albeit to a lesser degree.

Of the three towns, Jinja is the closest representative of the model of market and networks. As discussed in the previous section, this is because of the presence of economic actors in the SWM arena. It is a close representation because as will be argued below, these actors are not exactly the determining players in the field of SWM, yet they provide most of the SWM services to the council and its people.

Jinja has three administrative divisions: Central, Walukuba and Mpumudde. Due to the efforts to move towards privatisation in the country, the council contracted out solid waste management, specifically collection and transfer to the disposal site, through open bidding.

The arrangement is an annual contract between the municipal council and the private entrepreneur and in the last financial year (2008/2009) two contractors won the tenders. One serves two divisions: Central and Walukuba, while the other contractor provides service to Mpumudde. These contactors are actually private entities in form of companies with several employees and casuals (see the profile of one of the companies in Box 4.1). The casuals do the collection and sweeping. Payments from the council to the contractors are made as per the number of skips emptied to the disposal grounds. The contractor earns Ugshs 28,000 (USD 20) per small skip (3 tons) emptied and Ugshs 30,000 (USD 21.4) for a bigger skip (3.5 tons) emptied. On average the contractor serving the two divisions empties 18 skips per day while the one serving Mpumudde division empties on average 7 skips per day.

One outstanding aspect of the Jinja arrangement which is not the case for Mwanza and Kisumu (and which is also contrary to the 'markets' arrangement), is that households do not pay for the service. The reticence to pay for SWM is because households are convinced that it is the responsibility of the council to provide the service and at no cost. This conviction remains very strong so that attempts to introduce a fee of Ugshs 2000 (USD 1.42) per household per month were not successful. This is further depicted in the results of the survey which revealed the following percentages (See Table 4.1) about who the households perceived to be their service providers as relates to waste collection:

Interestingly, still 38.5% of the households interviewed perceive the council to be the one collecting their waste. Yet from the field work, it is very clear that the collection and disposal of solid waste from households is done by the two contractors (companies) awarded contracts by the council. A particular reason for explaining this surprising percentage could be that the contractor often moves around with the public health officer (who is an officer from the council) to identify which skips are full and need emptying. Besides, the contractors also use council vehicles branded 'Jinja Municipal Council' to collect the skips and this may appear as if the council is doing the

---

Box 4.1 Profile of Solid Waste Management Enterprise Limited.

The enterprise started operating in 2005 as a CBO but changed status to a private enterprise in 2007. It is owned by an individual-Beatrice Arigo. At the moment the company provides SWM services to Central and Walukuba Division. It employs 18 casuals who do collecting of waste and sweeping the streets. The company is only involved in SWM activities that is collecting and emptying skips and sweeping. They also collect plastics and metal waste and sell to recycling firms.

Table 4.1. Perceived service provider by households (n=218) in Jinja.

Service provider	Frequency	Percentage
CBO	1	0.5
Municipality	84	38.5
Private company	45	20.6
Others	88	40.4
Total	218	100.0

Source: Household survey in Jinja, 2009.

collection. It is therefore important for the council to have the capacity to rally the citizenry behind the established public and private partnerships even through civic education. This will help in sensitising the households on who provides them with SWM service and subsequently on the need to pay for it.

The important category (40%) referred to others, includes 38.5% of the households who responded that they provide their own service (self). These are households that manage their own solid waste either through burning or burying or re-using on farms. Finally there was a small group of housholds (1.9%) whose response was categorised as 'none.'

Networks

Even within the 'market model', the study established that there are still other actors with shared interests in SWM though the degree of cohesion varies between them. There is the involvement of women and youth groups in the road sweeping and clean-up activities which are done occasionally and mostly on a voluntary basis. There are also environmental groups that are actively involved, including NEMA as a wing of the government which has established pedagogic centres to showcase exemplary activities and is helping to source additional skips to be used in the council. This effort by NEMA augments the services provided by the contractors who are actually paid depending on the number of skips emptied, therefore newer and better quality skips is of added value for the work of the contractors. Also involved are international institutions like Lake Victoria Region Local Authorities Cooperation (LVRLAC) who promote exchange of practices amongst the councils member of the organisation. Also ILO and the Lake Victoria Basin Commission (LVBC) have actively taken part in capacity building. These efforts by different actors towards SWM not only depict the flexibility of SWM systems but through their different inputs they also ensure that the system remains sustainable.

Legitimacy and influence on decision-making

Legitimacy and influence on decision making acquire specific interpretation in a market model. Inherent in the formal authority given to contractors to operate, is the presence of a suitable environment for competition, speed, allowing greater room for market forces and making the use of commercial criteria appropriate. In the market sector, time is of essence. Thus timeliness in ratifying contracts, paying on schedule with minimal bureaucracy are critical for making a good return on an investment (Adei, 2009).

The legitimacy of the contractors is realised in a number of ways. First the Jinja Solid Waste Management By-Laws 2005, in their objectives recognise the role of private companies in the collection and disposal of waste, when this is practiced in a sustainable manner and at a fee. The two contractors serving currently are thus officially recognised by the council and in turn by a number of the clients they serve (20.6% in Table 4.1).

Secondly, the tenders for contracts in SWM are advertised through the media and as earlier mentioned, there is open bidding. Contracts for companies that were involved previously can only be renewed on the basis of their performance. Prior to the commencement of their work, they took part in an introduction training facilitated by the council and ILO. Payments to the contractors are made from the property tax accounts, indicating that they are factored-in in the council budget. This arrangement is however not free from complaints from the contractors, such as delayed payments which affect the robust nature of the system. Adei (2009) points out that such delays can only be prevented if the regulatory framework is changed in such a way as to make delays and inefficiency costly to an official (in this case, the council official in charge) in terms of promotion, remuneration and overall performance assessment.

As far as decision making is concerned, the council is still at the helm of SWM, making policies and seeing to their implementation. The contractors do not attend council meetings and are therefore unable to make or influence decision-making. From an interview with one of the contractors however, it became clear that they are free to voice their opinions directly to the town clerk, which may or may not be taken into account when formulating policies.

Relations and alliances

Here the focus is on the stability of the relationship between the municipality and the contractors and also amongst the contractors themselves. For this aim the study intended to find out whether this relationship is strong (based on a written agreement or contract) or loose (a simple supportive relation based on verbal agreement) (see Grafakos and Baud 1999). Furthermore, the relationship between the households and the contractors in this study is assessed by looking at their satisfaction about the service provided.

a. *With municipality.* As mentioned earlier, the arrangement in Jinja is a formal contract. The study found out that the contractor works together with the Divisional health inspector in identifying which skips are full and therefore need emptying. This exercise is done every day. There is therefore close monitoring of the waste-flows (a very relevant authority doing the monitoring on a very regular basis) and there is also a monetisation of the waste flow going on (contractors get paid for skips delivered). These practices are in themselves a first step in (ecologically) modernising the waste handling system. The arrangements are a mixture of market and state dynamics.

    In addition, the contractors use council vehicles for transporting the skips to the dumpsite. They hire the vehicles at Ugshs 100,000 (USD 71.43) per truck per month. They cover the costs for minor and major repairs and fuel as well. The drivers of the trucks are however employed by the council and not by the contractors themselves. This is to allow the municipality to control and monitor the disposal of waste because the contractor is paid as per the number of skips emptied. From the field interviews, this arrangement that does not seem to auger well with one of the contractors who said:

> I prefer to have my own driver(s), that way the I can give instructions accordingly and make my own schedule for waste collection, it is hard to dictate the council driver because I am not his employer.

b. *With other contractor.* As far as relations between the contractors themselves is concerned, it was noted during the field interview that market competition prevents the two contractors in Jinja from having any form of cooperation. As their contracts have an annual nature, they need to stay on top of the game to win the contract the coming year.

c. *With households.* Of the 130 households who responded to be receiving service (either perceived to result from the municipality, CBO or private sector) their rating on their satisfaction of service is indicated in Table 4.2. These figures show that about 62.6% (57.7% - satisfactory plus 4.6% - very satisfactory) of this groups responded to be satisfied with the services provided. In cases of filled-up and uncollected skips, households have been able to raise complaints through the local media or report directly to the health inspectors. From the council records, since privatisation waste collection in Jinja has gone up and varies between 40-60% of the total amount of waste generated.

An arrangement such as this one in Jinja, would in an ideal setting ensure that every household in the council receives SWM services because the contracts are awarded annually and depend on the performance of the contractors.

Table 4.2. Household perceptions (n=130) towards SWM services in Jinja.

Service rate	Frequency	Percentage
Satisfactory	75	57.7
Very satisfactory	6	4.6
Not satisfactory	49	37.7
Total	130	100.0

Source: Household survey in Jinja, 2009.

Payment systems

The contractors are paid by the council per skip disposed of, while households do not pay for waste management per se. The payments to the contractors are made from the property tax and ideally this is an efficient and effective way to fully cover all solid waste costs. In practice, however, there are challenges on the implementation of the property tax including issues related to the adequacy of cadastral property and the appraisal system and the efficiency of collecting the tax. This situation, makes payment recovery from all residents through property tax an un-feasible option.

A question asking the households to indicate which service (choosing between water, electricity, security and waste management) they would pay for first revealed that payment for waste management is not among their top priorities. (see Table 4.3). Only, 8.7% would pay for waste management as their first priority in service-delivery, while 52% percent indicated that they would pay for it as the last service amongst the four (see Annex 3 for details).

If households made payments (user fees), markets are also believed to empower citizens in the same way as we exercise powers as consumers. Introducing a market-like situation for public services allows clients to choose directly, hence the final say on public services rests with the people (Pierre *et al.*, 2000).

Introducing charges for waste collection is important because only a fee reflecting the costs will encourage users to value the service they receive correctly (McDonald and Pape, 2002). The fee can be introduced gradually. A practical way to do this may be to have a realistic flat rate with progressive increments. Given that it may prove difficult to measure and price household waste collection on volumetric basis (so that block tariffs can be applied like in water or electricity), equity concerns can be dealt with through the application of differential rates along income lines. This policy can be guided by some form of property valuation, that is the higher the value of the home or the rent one pays, the more one pays for waste collection. Strict regulations to ban burning or illegal disposal of waste should be formulated and enforced because currently this may provide an easier option compared to paying for waste collection. This preference is evident for 40.4% of the households (see Table 4.1) who responded to be managing their own waste and one of the ways they use is burning.

The risk with such a model (market arrangement) as discussed by Rakodi (2003) and also revealed by the study is that since it is market-driven with contracts renewed annually, it is likely to be short term oriented and fragmented. No investor would be willing to invest adequately when it remains uncertain whether they would win the contract again in the coming year or not. Therefore their activities are just oriented towards keeping the payments from the council coming but not to make progressive efforts to improve solid waste management as a whole. Currently the arrangement of the service provision is not guided by income levels but in the event that households

Table 4.3. Rating and priority of payment for services provision in Jinja (n=128).

Services	Frequency	Percentage
Electricity	12	5.5
Security	53	24.3
Waste management	19	8.7
Water	134	61.5
Total	218	100.0

Source: Household survey in Jinja, 2009.

are made to pay for the service, questions would arise with regard to the provision for low income households. As Prasad (2006) in Oosterveer (2009) suggests, there is a significant conflict between social development on the one hand and the private sector's motive for profit maximisation on the other. From the private sector's perspective, low-income areas are unattractive because of their limited accessibility and they have high levels of risks regarding non-payment.

4.3.2 Mwanza: governance as communities and networks

In the governance model of 'communities and networks', the communities are the major implementers of SWM with minimal state involvement. The networks aspect is relevant because of the participation of other actors in SWM with varying degrees of inclusion.

Mwanza's SWM governance arrangements comes close to that which can be described as 'communities and networks'. They come 'close' to this model because there are questions on extent to which these CBOs can exercise power. Although much of the SWM arena is dominated by CBOs and these CBOs are the major implementers of the policy, the council still dictates these policies, awards the contracts and generally steers everything that has to do with SWM.

There are 21 wards in Mwanza, out of which only 14 wards receive solid waste management services. These are the wards in the urban sections of Mwanza city. Privatisation of solid waste management resulted in the council awarding contracts to groups and in the financial year 2008/2009, contracts were awarded to Community Based Organisations and two private companies that serve the wards in the Central Business District. The incorporation of the different actors and leaving the CBD which is a prime area for the private companies is one step in the direction of a MM approach in Mwanza. Every other ward is served by one or two CBOs. The contracts are annual and, like in Jinja, this has a negative aspect because it does not encourage the contractors to improve on their SWM efforts as they do not know whether they will win the tender the following year or not. One contractor actually says:

> Why should I invest money in buying better equipment or even just making additional
> efforts as I may not win the tender next year, it would be better if the contract was
> for two or five years.

Considering a change in the contract arrangement to increase the length of the contract may resolve the dilemma above. The current contractual agreement is too short whereas the depreciation rate for waste collection vehicles, is for instance usually 4-10 years (Cointreau-Levine *et al.*, 2000). Contract arrangements that run for three to five years with performance based incentives that are payable in addition to the fixed fee, when the contractor meets or exceeds specified performance targets, will work to encourage the contractors to invest adequately in SWM (see Kennedy, 2002; Craythorne, 2006). The challenge here for the CBOs in particular would be to seek additional funds in order to win tenders.

The contractors in Mwanza charge different rates for different land uses but all households pay a standard fee of Tshs 400 (USD 0.28) per household per month. At the time of the field work, all the CBOs and the private companies interviewed stated that a number of households still do not comply to making the payments of the USD 0.28 because, among many other reasons, of the belief that it is the council's responsibility to provide the service. This lack of regular payment impacts negatively on the sustainability of the groups providing the services.

Like in Jinja, the household survey carried out revealed that 25% of the households still think that it is the council who manages their waste (see Table 4.4) despite the clear revelation from field work that this work has been fully contracted out to CBOs and private companies. Just like in Jinja, again, in Mwanza this could also be due to the assumption that since the council is in charge of waste management, it is also the one providing the service.

Table 4.4. Perceived service provider in Mwanza (n=200).

Service provider	Frequency	Percentage
CBO	88	44
Municipal	50	25
Private company.	27	13.5
Self	35	17.5
Total	200	100.0

Source: Household Survey in Mwanza, 2009.

Networks

Networks in the Mwanza arrangement are visible in the different actors involved in SWM albeit to different degrees. Apart from the council, CBOs and the private companies, just like in Jinja, there is NEMA, different government ministries and regional organisations, in particular, LVRLAC to which Mwanza is a member and the LVBC. ILO has been very instrumental in training the CBOs to earn their income from the waste collection and also urging them to form an association.

Legitimacy and influence on decision-making

Here, the emphasis is on legitimacy as a moral justification for political and social action. As Bratton (1989) says, it is a question of who has the right to assert leadership, to organise people, and to allocate resources in the development of enterprises.

First, the CBOs and the two private companies are legitimate organisations, officially recognised by the council and the people they serve. They are awarded formal contracts after having won through a democratic process (but not without complains of political interference). One public health inspector said:

> *While these contract arrangements are good, you cannot stop 'wazee' (the senior political officers) from influencing who gets a contract in a particular ward; there is a lot of pushing and pulling.*

The groups undergo registration as solid waste management service providers, pay a registration fee of Tshs 25,000 (USD 17.85). The private companies pay taxes to the Tanzania Revenue Authority.

Secondly these contractors are well known to the people they serve because the members of these CBOs are local and belong to/are residents in the wards they serve. Of the 82% who responded in the Table 4.4 as having known who their service provider is, about 70% are aware it is either a CBO or a private company. SWM provides a form of employment to these groups and in turn they are able to keep their surroundings clean.

The groups have been allocated wards as shown in the Table 4.5. During the field study, all sixteen groups were interviewed. They are groups made up of members ranging from six to as many as 30 (see the profile of one of the companies in Box 4.2). And it became clear that each contractor is to:
- sweep waste from the roads, drainage systems and open areas;
- remove sand on tarmac roads and drainage systems;
- unblock the drainage systems where there is dirt;
- collect waste and take it to transfer stations apart from the private companies who take their waste directly to the dumpsite;
- collect the fee from clients according to the Refuse Collection and Disposal By-law;
- ensure that the respective areas served are clean;
- prepare a programme/timetable for cleaning to be followed by the clients;

Table 4.5. CBOs/private companies and the wards served in Mwanza.

	Ward	CBO/private companies
1	Pamba	Prima Bins
2	Nyamagana	Ujasiliamali Cooperation Limited
3	Isamilo	Etia
4	Kirumba	Mwepe
5	Mirongo	Chassama
6	Mbugani	Maendeleo Mbugani
7	Nyakato A	Uzota
8	Nyakato B	Tufuma
9	Nyamanoro A	Muungano Wa Wajane
10	Nyamanoro B	Maendeleo Mkudi Kilimahewa
11	Pasiansi	Patuma
12	Igogo	Kinyagesi B
13	Mkuyuni	Himaja
14	Butimba	Boresha Mazingira
15	Igoma	Mkuwa
16	Kitangiri	Charity organisation

Source: Field work in Mwanza, 2007-2009.

> **Box 4.2. Example of a charity organisation profile.**
>
> The organisation serves in Kitangiri Ward. It started in 2002 with eleven members and today the group has grown to 25 members, 14 women and 11 men. Their main roles in SWM is waste collection from point of generation to the transfer station. Every household, hotel and street in Kitangiri is served by this CBO. Households pay USD 0.28 but for cleaning tarmac roads, the payment is done by the council. The CBO complained about delayed payments from council.

- submit to the council an outline proposal or action plan showing how the work will be done in the respective ward;
- submit to the council a list of clients every three months;
- submit a monthly report of work to the city head of public health;
- ensure workers have proper outfits and protective gears for work.

One unique feature of the Mwanza arrangement is that all households are to pay a standard fee of USD 0.28 per household per month. In the event of non-payments the contractors can seek redress from the council legal office. Legal action is however expensive and politically sensitive. This is especially the case when political elections are about to take place as no political aspirant would like to go against his/her voters interests. It is also challenging to follow up on households that do not pay and a contractor in the CBD puts it is this way:

> *The cost I would incur in taking up a legal case with a household that does not pay would be much more than the amount of service charge for waste collection that such a household is to pay, besides a tenant may move house, how do I even trace such a person.*

Even with such an arrangement of massive community involvement which, according to Pierre and Peters (2000), should result in minimum local government involvement and communities solving their own problems, the Mwanza city council is still at the helm of service provision, awarding contracts, and allocating roles and responsibilities.

As much as the groups are legitimate outfits with formal contracts and recognition by society, their role in SWM is weakened because of their minimal influence in decision making (see Helmsing, 2002) and the cases of non-payments by households who still perceive waste management as the responsibility of the council and of interference from political interests.

Relations and alliance

Just like in Jinja, also in Mwanza an assessment is made of the relationships between the different actors. While the relationship between the contractors and the municipality in Jinja is primarily based on sharing SWM responsibilities, in Mwanza, this is complemented by the sharing of the financial burden.

a. *With the municipality.* The community groups were asked how they communicate their needs to the council and all the 16 interviewed responded to be communicating through a public health officer representing the council in every ward. This officer monitors the SWM activities in the assigned ward and communicates the concerns of the CBOs to the city director. Furthermore when assessing relations and alliances, it emerged that every ward has at least two transfer stations either in the form of open grounds collection facilities or a skip. The location of the transfer stations is decided upon by the public and the CBOs. The council is responsible for collecting the solid waste from the different transfer stations to the disposal grounds. (This interplay between different actors in designating transfer stations location and even in transporting waste from the point of generation to transfer stations by CBOs and to disposal sites by the council is another step in line with the MMA). The CBOs pay Tshs 8,000 (USD 5.7) per trip to the council for transferring waste to the disposal grounds. The private companies however take their own waste to the disposal grounds and pay for its disposal. In addition, the CBOs and the private companies that have been awarded the SWM contracts are paid by the city council at a rate of USD 1.2 for every 300 m length of tarmac road that is cleaned daily.

On being asked if they receive any assistance from the council to further gauge their relationship with the council, all the groups, responded to be getting occasional legal assistance from the council to tackle defaulters of payments especially when it concerned business premises.

b. *With other contractors.* Unlike the situation in Jinja, as far as the contractors relating with each other is concerned, all the 16 groups interviewed in Mwanza belong to an association called the Mwanza Solid Waste Management Association (MASMA) (See the profile in Box 4.3). MASMA meets once every month to share ideas on problem solving and opportunities that can be explored further. Apart from the association, neighbouring CBOs (that is CBOs

---

Box 4.3. Profile of MASMA.

The association MASMA was formed in 2005 but officially registered in 2006. All the 16 CBOs involved in SWW in Mwanza are members of MASMA. The association charges Tshs 100,000 (USD 71.42) as registration fees for each CBO. The association seeks to pull together service providers with different levels of experience, knowledge and status, providing an enriched forum for experience and knowledge sharing, hence for improving the working performance of its members.

---

working in neighbouring wards) work together in sharing experiences and sometimes even the use of equipment in case the workload is more than expected.

c. *With households.* Household satisfaction in Mwanza is low compared to that in Jinja. Where n=165, only 51% indicated to be satisfied about the service provision. Possible reasons for this could be that the largest percentage of service providers are CBOs who, as it emerged, have no incentive to invest and improve SWM. This could impact on their level of professionalism. It is also possible that the population has very high levels of expectation about the performance of the contractors. A number of the households gave recommendations in line with improving the skills of CBOs, improving the infrastructure used for collection, showing that they expect more than they are receiving (see Annex 4).

In general though, privatisation with the involvement of communities in SWM in Mwanza has considerably improved the waste collection rate. Of the 296 tons of waste generated daily, 261 tons gets to the dumpsite, which is about 88% of waste generated. Before privatisation, only 28% of waste generated arrived at the dumpsite.

Payment systems

As mentioned before, households pay Tshs 400 (USD 0.28) for waste service per household per month. This is paid directly to the contractors and of the 165 households (82.5%) who responded to receive services either from municipality or CBO or private company, only a paltry 3.6% did not pay for these services. Among the households (n=200) 1% considered the fees too low, 45.5% considered them low, 32.5% moderate and 21% responded that it was not applicable or gave no answer. Interestingly though, SWM was not considered a priority compared to other services because in responding to the question their priority in payments for the different services, 76.5% (where n=200) indicated that they would pay for waste as the last service amongst the four (that is water, electricity, security and waste collection) while 88% would pay for water as a first (priority) service. There are many reasons for such findings, one of them could be as explained by McDonald and Pape (2002), who argue that receiving a service for free or having it heavily subsidised distorts not only its exchange value but its use value as well.

In the end of this analysis, although CBOs are perceived to be more participatory, less bureaucratic, more flexible, cost effective and having the ability to reach the poor and disadvantaged groups, using CBOs as the bulk service providers also brings to question issues regarding their sustainability. Some of the groups involved are small scale and depend on aid from other sectors in society. Non-payments by some households may contribute further to the unsustainability of CBOs operations. This situation has resulted in a reduction of the membership of some of the groups concerned and this means that less work can be done. Some of the groups had to diversify their activities, so now some are engaging in poultry keeping, stationery services (like photocopying and printing services) while others have sought loans from financial institutions to keep their activities running. This problem could possibly be solved by better informing the citizenry about the importance and practice of public-private partnerships. An active role by the council to facilitate payments by all households would be another contribution to the sustainability of these CBOs but gradual increments of the fee to be paid by the households would probably be more promising to the CBOs.

4.3.3 Kisumu: governance as hierarchy and networks

The governance model of 'hierarchies and networks' refers to governance conducted by and through vertically integrated state structures with the imposition of laws and other forms of regulation. The network aspect is relevant because of the presence of other actors in the domain of SWM.

The situation in Kisumu comes close to this model of hierarchy and networks. This is the case because, as is already presented earlier in this thesis, the council is still solely responsible for solid waste management and the management style is actually still of the command-and-control type. The Department of Environment receives its directions and authority from the line ministry of Local Government and implements them at the local level. Unlike Mwanza and Jinja, Kisumu has no formal/official arrangement that involves non-state actors in collecting and transporting waste or in sweeping the roads.

The council does the road sweeping itself, as well as the collection, transfer and disposal of waste, but these council services are concentrated in the Central Business District and only a few residential areas also benefit from them. Non-state actors provide service to most of the other residential areas in an unofficial manner.

Despite the dominant role of local government authorities there are however also networks, because a number of other actors take part in SWM including the informal/unofficial groupings[13], government ministries, NEMA, international organisations and others.

The field survey revealed the categorisation of the actors as presented in Table 4.6. These actors were involved in collecting, transporting, recycling and re-using solid waste in different residential areas.

Out of the 68 groups identified in Kisumu, 31 groups were interviewed. These groups have varying numbers of members ranging from 2 to 80. The profiles of three different groups that are involved in SWM in Kisumu are given in Box 4.4-4.6. Out of the 31 interviewed groups, 22 (71%) have been working on solid waste for more than five years meaning they were formed in the year 2005 or earlier. This duration of their existence speaks for their consistency. Those interviewed revealed they take part in different activities in solid waste management including:

Table 4.6. SWM Groups in Kisumu City Council and those interviewed.

SWM	Number existing	Number interviewed
Recyclers	23	6
Groups: CBOs and youth groups	27	17
Private companies/individuals	18	8
Total	68	31

Source: Field work in Kisumu, 2008-2009.

---

[13] Informal groupings stand for all those other non-state actors who are not officially recognized by the council as SWM actors though they are actively involved in SWM. They are listed in Table 5.4.

Box 4.4. Profile of Bamato environmental and sanitation project (recycler).

This is a self-help group that started in 2000 and was registered as an association in 2001. Today the group has 25 members and four satellite points for the collection of plastics. The group buys the waste plastic from scavengers, individuals and other private collectors and recycles this into products that can be sold.

Box 4.5. Profile of Carells Garbage Solution Company (private company).

The company was formed in 2005 and has two members. They collect garbage once a week from households and charge Ksh 150 per household per month (USD 2.14). The company provides waste bags to households and has a pick-up truck.

Box 4.6. Profile of Ten Stars-Tuungane (youth group).

The group was formed in 2007. Ten youth groups have come together and today they have grown to a total number of 60 members. They collect waste from two estates, once or twice a week for free. Their activities are financed by Tuungane youth project which has other income generating activities as well.

- collection of scrap metals and plastics;
- re-use of containers;
- composting waste from households waste;
- recycling plastics and metals from household waste;
- collecting household SW/transporting to transfer stations and dumpsite.

Legitimacy and influence on decision making

Even with the municipality as the central locus of authority, legitimacy remains a legitimate concern here to be able to ascertain the legal mandate accorded to the groups providing SWM informally. Therefore in addition to the moral and social dimensions mentioned under the Mwanza arrangement, the legal dimension is vital as well (see Vedder, 2003). Most of these groups are registered by the ministry in charge of community development, they are however not formally recognised by law as actors in the domain of solid waste management. The presence and activities of these groups are nevertheless known by the council and some even responded to be operating through some form of 'franchise'[14] in areas allocated to them by the council. A question posed to the groups to try to establish their legitimacy, resulted in responses provided in Table 4.7.

---

[14] The word franchise is in quotation marks because there are no legal papers to show for it and the arrangement is only franchise by name but not in actual sense.

Table 4.7. Legitimacy of groups.

Form of arrangement	Numbers
'Franchise'	11
Quasi contract	1
Partnership	1
Unwritten authority to operate	2
Pay rent to council	1
None	15
Total	31

Source: Field work in Kisumu, 2008-2009.

With no legal papers to show the arrangements they are part of, most if not all of these groups are not legitimate in the SWM arena. This impinges on a number of issues, for instance seeking legal redress in case of payment defaults becomes a problem. Getting donor assistance also becomes a problem because questions will arise as concerning ties to the public, transparency and adherence to the mission of a group, representative status and the relationship between the group and the community served. On the other hand, in terms of community support, openness of information, democratic decision-making, these groups can be considered more legitimate than some official actors are.

The groups engaged in waste services are composed of local members, some serving their own residential areas and others doing it at no cost. As it turned out from the household survey, these groups form the largest percentage of SWM service providers (37.5%), combining the percentages on CBO (10.5%) and private company (27%) (see Table 4.8). Responses that were categorised as 'Others' and which account for 53.5% of all responses (see Table 4.9) included responses like 'dumping in pits', 'burning' or 'the landlord comes to collect the waste'.

As far as decision making is concerned, like in Jinja and Mwanza, also in Kisumu the council remains at the helm of SWM activities. The groups mentioned do not attend council meetings nor influence decision-making in any other manner.

Table 4.8. Service providers to households (n=200) in Kisumu.

Service provider	Frequency	Percentage
CBO	21	10.5
Municipality	18	9.0
Private company	54	27.0
Others	107	53.5
Total	200	100.0

Source: Household survey in Kisumu.

Table 4.9. Means of communication between groups and council.

Means of communication	Number
Through Dept. of Environment	21
Through the NGO called Sana[1]	1
Through the association (Kiwama)	5
Through a broker	1
None	3
Total	31

Source: Field work in Kisumu, 2008-2009.

[1] Sana is a registered NGO in Kenya established to encourage water and sanitation development in the Nyanza region. It has supported a number of CBOs involved in solid waste management. See http://sanainternational.20m.com/About%20Sana.htm.1

Relations and alliances

a. *With the municipality.* The groups were asked how they communicate their needs to the council to assess whether there is any form of communication and therefore a relationship with the council. The responses to the question are summarised in Table 4.9. About 68% of all groups communicate through the Department of Environment which further indicates that the council is aware of their existence and that also their activities are known by the council. When asked if they receive any form of assistance from the council to further gauge the relationship they have with council, the groups gave responses which are listed in Table 4.10. These percentages clearly show that the council is, to a certain extent, working together with the different groups. The 35% mentioned under capacity building' refer to a number of the groups out of the 31 groups that have benefited from workshops and seminars facilitated by

Table 4.10. Assistance from the councils to the SWM groups.

Form of assistance	Percentage (where n=31)
Capacity building	35
Trucks for transportation	35
Networking	9.7
Dumping waste for free	3
Conflict resolution	3
Managing dumpsite	6
Helping to get fee from defaulters	6
None	19

Source: Field work in Kisumu, 2008-2009.

Shelter Forum, Practical Action and UN-Habitat under the auspices of the Council. These training sessions take place whenever funds are available and they have mainly been dealing with recycling opportunities, waste sorting and re-use.

b. *With other service providers.* In terms of relationships between the groups themselves, the groups revealed that they work together during cleanups and some even share their working equipment. Like in Mwanza, they also have formed an association called the Kisumu Waste Managers Association (KIWAMA; for a profile see Box 4.7).

   Of those interviewed 16% were also member of a scrap dealers association. This is different from the situation in Jinja where the two contractors do not work together, citing business competition as the reason.

c. *With the households.* When it comes to the relationships between the service providers and the households, the household survey revealed that 70.6% (n=93) of the households are satisfied with their service provision. In trying to understand this occurrence, a detailed analysis revealed that of those receiving waste collection services, 58% (n=93) receive these services from the private companies and that 87% of these (n=54) responded to being satisfied with service provision. This high appreciation of the services provided by the private companies largely explains the general satisfaction rate of 70.6%. Given that the private sector is driven by profit and ingenuity, which keeps it innovative and competitive, it comes as no surprise that a high percentage of those served expressed satisfaction in the service provided, as this is particularly the result of private sector involvement.

Box 4.7 KIWAMA.

The Kisumu Waste Managers Association (KIWAMA) was duly registered in 2009 at the Attorney General's office. The constitution governing its operations was enacted in the year 2009 and is a public document. KIWAMA also has a SACCO that operates as a micro-finance organisation, registered with the Ministry of Cooperative Development. Membership is both corporate and individual And the registration fee is Kshs 500 payable once upon registration. The youth represents 70%, women 23% and men 17% of the members. There are monthly meetings for members.

   Roles and mandate:

- to bargain for its members on strategic positions with the local authority and other development partners;
- to sensitise members on their role in the SWM cycle in a bid to improve service delivery within Kisumu Municipal Council;
- to fundraise for activities from the members for instance in clean ups and environment days;
- to sign MOUs for partnerships on behalf of the members;
- to train members on entrepreneurial skills to increase their marginal returns.

Payment systems

Kisumu's scenarios differ from the other two councils when it comes to the payment systems. The areas that are served by the municipality have their costs taken care of in the water bill and some of the households responded that they pay for waste as part of their house rent. The private companies are operating in open competition and work purely on a willing-buyer-willing-seller basis. From the survey, their services are mostly offered in high and middle income estates. Payments are made at the end of the month as per a verbal agreement with the household. CBOs operate mostly in middle and low income areas and they also charge fees agreed upon with each household.

Given the high number of informal operators, they charge various fees for waste collection but the average fees for different residential areas are revealed through the household survey in Table 4.11.

Out of the 93 households that receive service from either the municipality, private company or a CBO, 79.6% pay for it and a high percentage of them (69.7% where n=74) considers the fees they pay moderate.

In terms of priority amongst services, just like in Jinja and Mwanza, households responded to consider paying for waste collection as the lowest amongst the four different services earlier mentioned. Where n=200, 51% would pay for waste as the fourth and last service.

The undoing of this arrangement in Kisumu is that it lacks steering and integration which ought to be the role of government in a wider governance system but on the other hand as argued by Rakodi, in a system in which municipal capacity is almost lacking, resources inadequate and bureaucratic processes lack efficacy, it is important for residents and businesses to cultivate not only political but also bureaucratic relationship - porous bureaucracy, the informal exchange relations between clients at all levels which ultimately may provide some of the poor with channels for obtaining access to services as personal or group favours. An example is that of *Manyatta* – a low income residential estate in Kisumu – where ten youth groups have come together and formed one group. They provide waste collection services to the residents at no fee. This complements the activities of other private collectors who provide service to some of the *Manyatta* residents at a fee.

4.4 Market, communities or hierarchical arrangement?

As mentioned at the beginning of this chapter, some observers suggest strengthening governmental authorities to take their responsibilities more seriously, whereas others consider the current African

Table 4.11. Payment Rates for SWM services.

Residential area	Payment rates (Kshs/month)
Low income areas	40.00-100.00
Middle income areas	150.00-250.00
High income areas	250.00-500.00

Source: Household Field Survey in Kisumu, 2009.

state fundamentally incapable of delivering such services in a reliable manner. According to the second view, active involvement of private companies, non-governmental organisations and local communities is pivotal to implement and manage urban environmental infrastructures.

The field studies reveal that in all the three municipal councils, the local government is still at the helm of SWM so it is not a full market-based governance for Jinja neither is it a complete community-based governance for Mwanza. Governments in the three case studies still enjoy an unrivalled position in society and they are still the obvious loci of political power and authority. Though they have been engaged in some kind of negotiation with other significant actors in society, their dominant role remains unquestioned. These local governments may become increasingly dependent on these other societal actors but they have remained in control of some unique power bases in society such as legislative powers, powers to award contracts and even the authority to determine service charges. Therefore while in theory, network governance could be the most appropriate model to manage SWM, the realities on the ground echo the need for (and perhaps a renewed interest in) recognising the importance of an active state in managing SWM. Given that an active engagement and effectiveness of the state remain critical variables, the state (with a focus on the municipalities who are charged with SWM) has to acquire skills which in the past were associated with the private sector: strategic planning and management, effective time management, respect of contracts, and timely delivery of licenses and approvals among others. The state must have the capacity to engage in partnerships with non-state actors in joint ventures. The state must also have the capacity to transform the laws, rules and regulations made in an era when the state was looked upon as the sole provider of public goods and even in charge of organising businesses. Therefore as much as effective market economies and societal institutions are essential they require a functioning and capable state in order to operate and grow. At the same time, evidence for the active participation of non-state actors in SWM in the three towns is unquestionable and this calls for considering their position as argued in the network governance discourse in future governance arrangements. With that understanding and based on the criteria of Modernised Mixtures (flexibility, accessibility and sustainability), then for each town (whose solid waste status is summarised in Table 4.12), the following conclusions can be drawn.

The 'community dominated' arrangement in Mwanza which includes CBOs, private companies and the municipality is more flexible than the other two urban centres. The private companies cover the wards in the CBD as they have the resources (financial and human) to take on the bigger work load. Further, the CBD areas are well structured and can easily be reached. Returns on investment is likely higher in the CBD than in the wards away from the CBD, this makes it easy to contract out the CBD to the companies and have more flexible arrangements in the neighbourhoods. The CBOs cover the other wards and their capacity compared to the private companies is lower. The municipality does the supervision and monitoring. This arrangement has ensured the accessibility of SWM services by most if not all households as is shown in the number of households receiving these (82.5%). While 96.4% of those who receive these services actually pay for them, the level of satisfaction amongst them was low compared to the other two towns. This brings to question the issue of sustainability vis-à-vis the CBOs' large coverage. The low satisfaction in Mwanza compared to the other two urban centres is linked to the fact that CBOs who provide service to most of the households are not well equipped. Their contracts are annual, and this makes it not interesting for the CBOs to invest in their SWM equipment. The reason for using annual contracts is the

Table 4.12. Summation of SWM indicators in the three urban centres.

Town/SWM status	Jinja	Mwanza	Kisumu
Waste collection arrangement	Private collectors formally contracted and the municipality	CBOs and 2 private collectors formally contracted and the municipality	CBOs, private companies and operating informally and the municipality
Percentage of households receiving SWM service	60% (n=218)	82.5% (n=200)	46.5% (n= 200)
Percentage of households that pay for waste collection	N/A	96.4% (n=165)	79.4% (n=93)
Satisfaction of SWM amongst households	62.6% (n=130)	51% (n=165)	70.6% (n=93)
Percentage of waste collected in the towns	40-60%	88%	35-45% (municipal and non-municipal)

Source: based on field work 2007-2009.

municipality's aim to gauge performance and to only renew contracts on the basis of previous performance. For this arrangement to be sustainable though, the council should improve the working environment for these CBOs. The contract period could be made longer, for instance 5 years, and the council could be allowed to terminate the contract before the end of the five years' term if the contractor does not perform satisfactory. Performance based incentives could also be used to encourage the CBOs to deliver more. The user fees charged for collection should be gradually increased as well. The CBOs can then access loans from the existing microfinance institutions and in turn satisfy their clients better.

The 'market dominated' arrangement in Jinja having included private contractors, has reached a level of 60% in terms of service accessibility. The percentage can be increased further by bringing on board groups like CBOs to provide service to those who are not covered yet. According to Adei (2009) contractors cannot provide services below cost or without an acceptable minimum return on their capital and expertise. If that cannot be charged to consumers directly then other means have to be adopted. While today the contractors are paid from the property tax, their revenue can be improved by progressively demanding households to pay for the services provided. The Jinja municipal council can perhaps learn from Kampala City Council in Uganda which has awarded, via competitive tendering, a monopoly for a restricted area to (mostly local) private firms but subsidises firms working in poor areas (Oosterveer, 2009). This policy makes it somewhat affordable for poor households to also profit from SWM services. Privatisation is not to be excluded but should be firmly controlled (against corruption and underperformance) and not be limited to allowing contracts to large (foreign) companies, but include smaller local companies and NGOs/CBOs. Proponents of the neo-developmental state claim that it is only through active government interventions that access of the poor to environmental infrastructures can be secured, as their economic and political power is too limited to realise this otherwise.

The hierarchical arrangement in Kisumu ranks low when it comes to service accessibility and percentage of waste collected. This is because the municipal authority which is seen as the sole provider of SWM is limited in its resources and has not wholly linked up with the informal/ official sector. Though the percentage of households that receive SWM services (46.5%) is low compared to Jinja and Mwanza, the informal sector has certainly scored as is evident in percentage of service satisfaction. Therefore, there is need to rethink the official definition of informality and consider modernising the informal sector. Legitimising the involvement of these informal actors is one step towards improving the accessibility and sustainability of SWM. Providing an enabling environment in terms of well-functioning communication channels, adequate fee levels and active collaboration with municipalities would be another.

To this end, the definitive answer to how the urban poor can best be served is not community dominance, neither market dominance, nor hierarchical arrangements. A balanced arrangement is needed where all societal actors can play their role. It is clear that involving non-state actors as in a network governance arrangement is truly plausible and these actors, both formal and informal, need to work under an effective and strengthened government in order to afford all income groups to access solid waste management services and to ensure flexibility and sustainability of the services provided.

Chapter 5.
Opportunities for inter-municipal cooperation

5.1 Introduction

The previous chapters have covered governance arrangements surrounding solid waste management in Kisumu, Jinja and Mwanza which come second after the capital cities in terms of their urban status and size. Building on the findings in Kisumu (Kenya), the aim of this chapter is to assess the opportunities for inter-municipal cooperation in solid waste management amongst three smaller urban centres in Kenya found in the Lake Victoria Basin: Kisii, Homabay and Migori municipalities.

The location of the three councils (Kisii, Homabay and Migori) of the Lake Victoria Basin in Kenya is shown in the section boxed with the white line in Figure 5.1. More details on their location are presented in Figure 5.2. Figure 5.2 highlights the location and the distance between the three towns, while also some town councils are included that lie in their midst.

This chapter combines a discussion on SWM in the three small towns with a review of multi-level governance and inter-municipal cooperation related to solid waste. In section 5.2, the concept

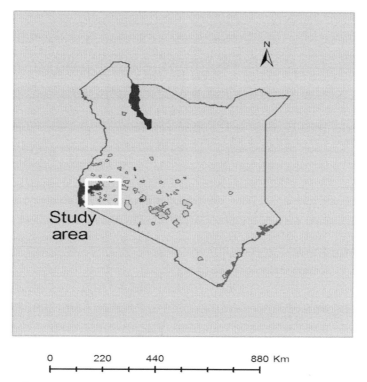

Figure 5.1. Map of Kenya showing a portion of Lake Victoria Basin (GeoTASADA Ltd, Kenya - May 2009).

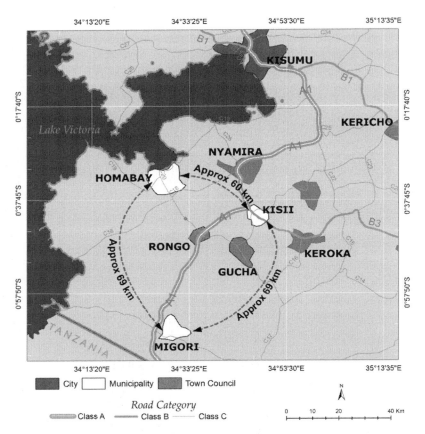

Figure 5.2. Part of Lake Victoria Basin in Kenya showing the three towns under study (GeoTASADA Ltd, Kenya - May 2009).

of multi-level governance is introduced and different organisational models for cooperation are discussed. Section 5.3 introduces the three towns within the broader research context of solid waste management pointing at opportunities of inter-municipal cooperation along the waste chain. Section 5.4 presents possible models for managing waste in the three towns while Section 5.5 summarises the findings by drawing parallels between multi-level governance and the modernised mixtures approach. Section 5.6 present the conclusion.

5.2 Multi-level governance

This study seeks to discuss the multi-level governance (MLG) perspective by focusing on inter-municipal cooperation framed along the arguments of what Marks and Hooghe (2003) refer to as the 'Type II form of MLG'.

    While the Type I form of MLG builds upon 'general-purpose' jurisdictions (governments) at different levels, and is mostly interested in the interactions between these levels and the sharing of competences between them, Type II MLG is characterised by task-specific (instead of general purpose) jurisdictions, intersecting memberships and a flexible design that is responsive to temporary needs (Marks and Hooghe, 2003). Type II MLG consists of special-purpose jurisdictions that tailor membership, rules of operation, and functions to a particular policy problem. Such jurisdictions may very well span several (territorially or otherwise defined) levels. Provision is made for interactive arrangements in which public as well as private actors participate, aimed at solving societal problems or creating social opportunities.

Kenya has only two levels (tiers) of government - local and central/national with no regional level government in between them. This permits the study to take on MLG II which is not focused on nested administrative entities and therefore allows to move away from the hierarchical state centric arrangements as provided by Type I MLG.

The Lake Victoria Basin offers the functional space, or territorial arena, within which the three municipalities can experience territorial interdependence and geographical proximity which are important motivating factors for working together to solve shared problems or discover opportunities for engaging with the outside world (Conzelmann, 2008).

5.2.1 Multi-level governance and improved solid waste management

As managing solid waste at the municipal level has grown increasingly complex in terms of equipment, technology, personnel and the associated costs, rural and small councils are likely to find it difficult to fulfil their responsibilities in this area. Unlike their larger urban counterparts, rural and small urban councils often have a lower tax base, which means that the revenues for financing waste management activities are limited. By working together, however, most of the activities in the waste chain including effective recycling programs (for instance, the marketing of recyclables and purchasing of goods with recycled content), putting up state-of-the-art landfills, and providing waste-to-energy facilities could be within the reach of even small communities with few resources. When councils combine financial, administrative, personnel, and equipment resources (EPA, 1994), the costs of pursuing certain projects is spread among several jurisdictions. Still, differences in available resources between participating councils must be considered when trying to achieve equity even as economies of scale are realised.

With multi-level governance, externalities can be internalised through inter-municipal cooperation, but conflicts may still occur, for instance, hauling waste across jurisdictions can cause conflicts. Cooperation sometimes can require that waste be transported over long distances and through neighbouring municipalities. Communities along the routes leading to a shared solid waste facility might witness an increase in traffic. Concerns over the resulting congestion, pollution, and wear and tear of the roads could create conflicts among the communities concerned. This could be addressed for instance, by restricting traffic to use certain roads or paying taxes to the affected councils or other voluntary agreements.

Multi-level governance ensures that decision-making is carried out at the level of government that is closest to the individual citizen – a concept referred to as the subsidiarity principle. Local responsiveness is also achieved because local officials are in a better position to respond to local tastes and preferences than are officials of senior levels of government. MLG in the form of inter-municipal cooperation also allows the different councils involved to retain a certain degree of autonomy depending on the organisational model settled upon. Voluntary cooperation for instance, is common where local autonomy is highly valued: municipalities can retain independence while reaping the benefits of cooperation (Conzelmann, 2008).

Multi-level governance allows division of responsibilities according to capacity and availability of resources at different levels of government. There are functions in the waste chain like waste collection that would be better managed by individual councils in-house, while others would be more effectively handled when councils cooperate. The retention of certain functions within each council allows the public greater access to local decision-making and better accountability on the implementation of these decisions.

Therefore possible benchmarks that could present themselves as the different aspects of SWM when considering MLG and inter-municipal cooperation, to be discussed in the empirical section include:
- economies of scale;
- equity;
- externalities;
- the subsidiarity principle; and
- accountability.

5.2.2 Organisational models for cooperation

Cooperation can take different forms and different authors have categorised them differently. Table 5.1 links the structures by drawing similarities from the explanations given by different authors/institution. There are plausible arguments in each category; but employing arguments from the Modernised Mixtures approach on which this study is based, would lead to the following concerns related to these organisation models. These models should:
- Reflect the local realities including the available resources amongst the participating councils, the institutional framework and the legislation in place.
- Consider the balance between technical and institutional concerns, that is, which institutional model is fit for which technical undertaking. Balance the economic and institutional concerns and also the economic and technical concerns.

These concerns come out clearly in EPAs categorisation in which the concept of regionalisation is prominent and the management structures are defined accordingly. Dollery and Johnson's (2005) categorisation also brings out the multi-level governance concept particularly in the first five models. These two categories are therefore taken into consideration in the rest of this study.

EPA (1994) uses the concept of regionalisation and distinguishes five management options for achieving this while giving strengths and weaknesses for each. They include:
- *Intergovernmental agreements*: these agreements are contracts between two or more municipalities to perform a specific task together. They may be informal arrangements or more complicated legal contracts. They constitute the most flexible model for regional cooperation, as councils can structure each project individually. On the other hand though, in the absence of a more formal organisation, financing for such projects can be difficult to obtain. Each participating council would have to raise money for the project individually. For

Table 5.1. Organisational models for cooperation.

Organisation model	EPA, 1994	Oakerson, 1999	Dollery and Johnson, 2005	Price Waterhouse Coopers, 2006
Category A	Intergovernmental agreements	Co-ordinated production Joint production	Voluntary arrangement Joint Boards	Cooperative cross council efforts
Category B				Shared corporate services
Category C	Commercial sector (franchising and contracting	Private contracting Franchising Vouchering		Specialised lead service provider
Category D	Regional councils		Regional organisation of councils	
Category E	Authorities/trusts/ districts			
Category F	Non-profit public corporation			
Category G			Agency model	
Category H			Amalgamation	

each new decision, all jurisdictions must regroup and reach a new agreement. Consequently, intergovernmental agreements often are better suited for more limited regional projects than for permanent ones.

- *Authorities, trusts and special districts*: they have political and financial autonomy. They are able to raise funds through bonds or taxes. Their autonomy helps these organisations sustain cooperative partnerships among communities and execute projects in an environment free of individual community politics. Member councils set up an advisory board or establish a reporting structure for the organisation that ensures proper oversight.
- *Regional councils*: sometimes referred to as councils of government, regional planning commissions, or regional development centres. A key characteristic of a regional council is its flexibility. They can be structured in such a way that they are able to meet the needs of the member councils. Through the council, public and private decision-makers can be brought together to consider a regional strategy. If regionalisation seems promising, the council then can plan and implement the program. Existing councils can lend their experience and enhance a shared sense of cooperation between different regional councils.
- *Non-profit public corporations*: they have financial autonomy and are able to issue tax-exempt bonds, making fundraising easier. Councils have control over decision-making because local officials sit on the board of directors. On the other hand, it can take a long time to establish such a public corporation because of the legal procedures involved and because political

considerations can influence the project, since councils/communities have control over decision-making.

- *Commercial sector*: contracting and franchising are the two most prevalent forms of commercial involvement in solid waste management services. As an organisational model, it can offer experience, access to state-of-the-art technologies, and lower costs. It can, however, also entail lengthy bidding procedures and require complicated contract negotiations. It may also mean that the participating councils have less control over the daily activities.

The model chosen for regional cooperation at any one time depends on factors as available financing, applicable laws, and existing government bodies or regional organisations. Another important factor in this regard is the amount of control that communities want to keep over the organisation and the type of SWM services and projects that the organisation oversees. For example, if a regionalisation effort entails constructing waste management facilities or providing solid waste services, a formal, legal structure with financing capabilities might be needed. For a one-time project or a limited, clearly defined effort (such as organising a hazardous household waste collection program or arranging for equipment sharing), a more flexible model like an intergovernmental/inter-councils agreement might be more appropriate.

There is also a taxonomic classification of alternative models of municipal governance developed by Dollery and Johnson (2005). They include:

a. *Voluntary arrangements* between geographically adjacent councils to share resources on an *ad hoc* basis whenever and wherever the perceived need arises.

b. *Regional organisations of councils* (ROCs) which constitute a formalisation of the *ad hoc* resource sharing model, typically financed by a fee levied on each member council as well as a pro rata contribution based on income rate, population, or some other proxy for size, which provides shared services to member councils.

c. *Area integration* or *joint board* models which retain autonomous existing councils with their current boundaries, but create a shared administration overseen by a joint board of elected councillors.

d. *Virtual local government*-consisting of several small adjacent 'virtual' councils with a common administrative structure or 'shared service centre'. Such a centre would provide the necessary administrative capacity to implement the policies decided upon by the individual councils, with service delivery itself contracted out either to private companies or to the shared service centre depending on the relative costs of service provision and the feasibility of using private firms

e. The *agency model* in which all service functions are run by state government agencies with state government funds and state government employees in the same way as state police forces or state emergency services presently operate. Elected councils would act as advisory bodies to these state agencies charged with determining the specific mix of services over their particular geographical jurisdictions.

f. The most extreme form occurs when several small councils are *amalgamated* into a single large municipality

5.2.3 Methodology

This part of the study employs an embedded case study research design explained already in Chapter 2. The data were collected in the following ways:

a. Document and literature review: documents with discussions on multi-level governance and inter-municipal cooperation in different parts of the world were reviewed. Next, different pieces of legislation in Kenya were also reviewed to find possible models for cooperation.

b. Direct observation was also used as a method of data collection particularly to observe the status of the waste management infrastructure in the three towns.

c. Given that inter-municipal cooperation in service provision is a relatively new concept in East Africa and in Kenya in particular, semi-structured interviews were used to obtain as additional information. The following resource persons were interviewed:
    * the town clerk in Kisii;
    * 4 public health officers- 2 in Kisii, and 2 in Migori;
    * two private collectors in Kisii;
    * two recycling firms in Kisii;
    * town engineer in Homabay in charge of SWM;
    * an NGO representative from Enviro-Watch Homabay;
    * project officer-Kenya's chapter working with LVRLAC.

d. Two stakeholders workshops were organised in Kisumu and Bukoba (Tanzania). The workshop in Kisumu was intended to discuss the possibility of working together as councils in the region. Some specific issues discussed here were the practical challenges and opportunities of inter-municipal cooperation: The workshop brought together 19 stakeholders:
    * Given that there are Multi-Stakeholder forums (MSF) in each council, the study used the MSF ideology in such a way that participants were picked from the different sections to make the workshop quorum representative. Three from each of the small urban centres were selected (one representative from the national government, one from the local government and one from the non-state organisations) to make 9 participants for the workshop.
    * Five participants from the Kisumu City Council to share their best practices (the director, deputy-director, two Councilors and one officer, all from the department of environment).
    * One representative from LVRLAC as a regional organisation.
    * Four researchers from the PROVIDE project.
    In the workshop in Bukoba, the coverage of stakeholders was wider and they came not only from Kenya, but also from Uganda and Tanzania. This workshop was the follow up to the Kisumu workshop and was organised under the auspices of LVRLAC. The output of this workshop is discussed further in Chapter 7.

e. To measure the distance between the three towns, first a map of classified roads for the study area was obtained and from this the inter-local authority routes were identified. The distance between the most central (mean centre) place in each local authority boundary was estimated by tracing the dimensions along the routes earlier identified. Spatial data for estimating the proximity of the municipalities were obtained from the digitised Survey of Kenya (SoK), with 1:50,000 toposheets. Data for the municipal/local authority boundaries were obtained from the ILRI (International Livestock Research Centre) GIS Unit. Using Arsis 9.3, centroids (mean centres) for the local authority polygons were derived. The distance from the centroids of target local authorities were then estimated by tracing along the interlinking road network using the measure tool in Arc-map. The rationale here was that solid waste transportation would most likely be undertaken by vehicular transport.

5.3 Solid waste management in the three small towns: opportunities for inter-municipal cooperation

5.3.1 Geographical location

Kisii, Homabay and Migori are all small urban centres found in Kenya in the Lake Victoria catchment and located within less than 100 kms from each other. Figure 5.2 gives a visual representation of the councils showing the distances between them by road and their location compared to Lake Victoria. Also included in the map are some of the town and county councils neighbouring the three councils who are likely to be affected by any cooperation efforts between Kisii, Migori and Homabay. Their ecological positioning is a first step towards inter-municipal cooperation. The councils will be able to internalise externalities of any joint activity but such consequences for the smaller town councils located in their midst need to be taken into consideration as well. The size and population figures of the three councils are presented in Table 5.2 and in combination with the level of service provision (see Table 5.3), these findings justify determining these municipalities as 'small'.

5.3.2 Institutional organisation

Activities of solid waste management in the three councils just like in the primary towns discussed earlier, are the responsibility of the local authorities. Inter-municipal cooperation would uphold this subsidiarity principle as the local authorities would still remain responsible for SWM. This involvement of municipal authorities further ensures local responsiveness, because these

Table 5.2. Municipalities' profiles.

Municipality	Kisii	Homabay	Migori
Size in km^2	29	197 (of which 50% is located in the lake)	58.4
Population			
1999[a]	65,253	56,297	95,446
2008 (projected)[b]	77,983	71,552	123,541

Source: Compiled by author using data from the councils collected in 2009.

[a] 1999 population figures from population census of 1999 (CBS).

[b] 2008 Population projections are based on information from district development plans and Central Bureau of Statistics

Table 5.3. Brief on SWM in the three municipalities.

Municipality	Kisii	Homabay	Migori
SWM institutional organisation	Under the public health department	Under the town engineer	Under the public health department
Staff numbers	54	49	50
Staff	1 public health officer 1 environmental officer 2 foremen 4 headmen 4 drivers 42 street cleaners	1 engineer 1 works officer 1 foreman 1 headman 2 drivers 4 loaders 39 sweepers	1 public health technician 1 supervisor 2 foremen 2 drivers 29 loaders 17 street cleaners
Waste generation	domestic, commercial, street sweeping, medical, institutional and industrial	domestic, commercial, street sweeping, medical, institutional	domestic, commercial, street sweeping, medical, institutional
Storage	litter bins, polythene bags, crude dumping points	litter bins, 4 transfer stations and crude dumping points	dumping points (hot spots) along the road
Collection	manual loading from dumping points	manual loading from transfer stations and dumping points	manual loading from dumping points
Frequency of collection	CBD daily by council main market daily daraja mbili market twice daily by council a few residential areas daily by youth groups other places occasionally by the council if the waste becomes a nuisance	CBD daily by council bus park daily by council Sofia market weekly by council main market twice daily by council other places occasionally by the council if the waste becomes a nuisance	CBD daily by council market daily bus park daily Along the main road (Kisii-Migori) from residential areas daily other places occasionally by the council if the waste becomes a nuisance
Transport equipment	1 tipper, 1 tractor	1 tipper, 1 tractor	1 tractor
Disposal site	about 1 km from CBD, next to a hospital and a river	about 1 km (within town residential area), no attendant, not fenced	about 5 kms from CBD, no attendant, not fenced
Private involvement	some youth groups in waste collection	some CBOs involved in street sweeping	none

Source: Author's construction from field interviews and councils' records, 2009.

local authorities are considered to be closest to the public as opposed to a national or regional government. Though the Local Government Act Cap 265 allows these local authorities to have joint operations, it does not specify which organisational models these authorities can take on. Interviews conducted in each council with officers responsible for waste management revealed that the three councils had never considered inter-municipal cooperation to provide solid waste management service. The officers gave a number of reasons why this is the case and this included: the different resource allocations for different activities in each council and the desire to maintain that autonomous status in running their services which according to them would be eroded by such cooperation. One public health officer in Kisii responded to the question on cooperation by asking;

> *Who would allow us to take our waste to their jurisdiction and why should we allow others to bring their waste to our town?*

These sentiments seem to have been fuelled partly by indictments over poor performance by small councils in Nyanza Province by the government because the field interviews were conducted not long after the following press report was released:

> *A new performance report released by the Government indicts civic authorities in Nyanza over poor performance. The province has produced eight of the ten worst performing local authorities. Besides, the region's 37 local authorities did not feature in the best ten among Kenya's 175 local authorities. According to the evaluation report released last month by President Kibaki, the poor performers are the town councils of Keroka, Nyamache, Masimba, Yala, Ogembo and Suneka. Others were Homabay County Council and Migori Municipal Council. According to experts, the poor showing is a result of the authorities' small sizes. Ten out of the 27 civic authorities in the region are categorised as 'small,' hence their limited viability'. Government Indicts Councils over Performance by Mangoa Mosota, Standard Newspaper 23rd June 2008.*

The three towns under study are all in Nyanza province and this poor performance had led to talks of merging such small councils.[15] The stakeholder workshop held in Kisumu that brought the different councils official from the three councils together however, provided room for discussions which allowed them to understand both the opportunities and challenges of inter-municipal cooperation in SWM.

For a long time, SWM in these councils has not been prioritised and has been reduced to just one of the several functions under the public health department as in Kisii and Migori or the engineers department as in Homabay. The impact of this situation has been that appropriate allocations for personnel, budget and finances, vehicles and equipment have received low priority while other 'more important' departments were allocated more resources. For instance, in Migori municipality, in the financial year 2007-2008, their Local Authority Transfer fund allocation was Kshs 26,670,627 (USD 381,009) (see Annex 5 for LATF allocation to the three councils over the

---

[15] Political Pride is no Reason to Turn Villages into Districts, Mike Owuor, Local Daily Newspaper, The Standard Edition 20th March 2009.

years). Their total expenditures on waste management were Kshs 2,937, 651 (USD 41,966), out of which only Kshs 339,000 (USD 4,843) was used on equipment and Kshs 150,000 (USD 2,143) on maintenance, the rest (83.4%) was used for worker's remuneration. The revenue from waste was only a paltry Kshs 257,000 (USD 3,671) as compared to the figure of expenditures. Establishing the Department of Environment in each council with its own budgetary allocation and seeing to it that the funds generated by the department are ring fenced for operation and maintenance will probably empower these councils in seeking solutions to SWM even beyond their jurisdictions.

Equity concerns

The three local authorities receive almost similar allocations from the Local Authority Transfer Fund (LATF). In the Financial Year 2008/2009, they got the following allocations:
- Migori   USD 427,158
- Homabay   USD 456,551
- Kisii   USD 409,572

These figures (and even those for previous years) do not show huge differences, which means that there would be no major equity concerns when it comes to cooperation amongst them. There are no clear rich or poor councils. The LATF is a reflection of the service provision and in turn the revenue because it is disbursed to improve service delivery and financial management, and to reduce the outstanding debt of the local authorities. At least 7% of the total fund is shared equally among the country's 175 local authorities; 60% of the fund is disbursed according to the relative population size of the local authorities. The balance is shared out based on the relative urban population densities. The likely challenge here is the amount that each council allocates to waste management.

5.3.3 Waste generation

Substantial solid waste is generated daily in these three municipalities. The typical municipal waste stream contains general waste (organics and recyclables), special wastes (household hazardous, medical and industrial waste), and construction and demolition debris. The waste stream is mixed because there is no separation or waste sorting done at the point of waste generation.

According to the field interviews, the generation rates are approximately 0.5 kgs per person per day (see report by CAS Consultants, 2005). This gives the following estimations of the quantity of waste generated per day in the year 2008:
- Kisii   39 tons
- Homabay   36 tons
- Migori   62 tons

The figures are estimated to increase together with the population growth (see Table 5.4). While these amounts may seem modest compared to the 1-2 kgs per person per day generated in developed countries, the problem is that most wastes in these councils is not collected through the municipal collection systems because of among other reasons inadequate waste management

Table 5.4. Population projected versus waste generated per day.

Municipalities	Population growth rate[1]	2008		2010		2015	
		pop.	tons/day	pop.	tons/day	pop.	tons/day
Kisii	2.0%	77,983	39	81,134	41	89,554	45
Homa-bay	2.7%	71,552	36	75,468	38	86,221	43
Migori	2.9%	123,541	62	130,810	65	150,910	76
Total		273,076	137	287412	144	326,685	164

[1] Population growth rates are taken from the Central Bureau of Statistics in Kenya based on 1999 population census.

resources. To be noted is that, over 70% of the waste generated in the councils is organic and this can be explained by their geographical location and economic activities. These towns are located within a very rich agricultural hinterland especially Kisii town. Fishing is a main income earner for the inhabitants of Homabay, while agriculture is also the major income earner in Migori. These agricultural and fishing activities generate considerable organic waste.

5.3.4 Collection and transportation

For health reasons, given that all three municipalities are in equatorial climates (highland equatorial for Kisii, inland equatorial for Homabay and mild inland equatorial for Migori) which means they witness relatively high temperatures, waste should actually be collected daily. This consideration makes the challenges and costs of solid waste management in these towns even more daunting. Moreover, because when services are available they generally are provided to the CBD and the wealthier neighbourhoods. In other areas, uncollected wastes accumulate at roadsides, are burned by the residents, or are disposed of in illegal dumps which leads to environmental pollution and increases the risks of disease outbreaks.

In all three municipalities, the solid waste generation exceeds the existing collection capacity. On average the amount collected does not go beyond 30% of the total quantity of waste generated for each town. For instance, Kisii Municipal Council has only one tipper and one tractor with the following dimensions (see Table 5.3):

- A Ford Model 5030 tractor with a non-tipping trailer. Trailer body dimensions 3.2×1.9×0.6 m heaped = 3.6 m^3.
- An Isuzu tipping truck. Truck body dimensions 3.2×2.3×0.9 m heaped = 6.6 m^3.

The tractor/trailer collects four loads of waste generated per day (five days per week) while the truck collects three loads per day (five days per week) and deliver them to the disposal site. Based on Manus (2005), the average density of waste in this town is 450 kg/m^3 (500 kg/m^3 from the market and 400 kg/m^3 from other transfer points). Calculating the total amount of waste that the two vehicles are able to transport to the dumpsite per day combining this information on the dimensions of the vehicles, the density of the waste and the loads per day would give:

- 8.91 tons/day for the truck;
- 6.48 tons/day for the tractor/trailer.

In total only 15.39 tons/days out of the 39 tons/day generated in Kisii ends up in the dumping site. It is generally considered that a collection rate of 80% provides a good service as it is estimated that 20% will be removed from the waste stream by recycling or composting. This means that whether the council retains the use of the existing models of vehicles or modifies them to include a skip trailer for example, there would still be a need to increase their numbers by 3 to 5 more (that includes a spare in case of breakdown) to efficiently manage the increasing quantities of waste.

As of July 2009, with the help of UN-Habitat, Homabay Council had acquired three 60 hp tractors with trailers for waste transportation, and a similar gesture was to be extended to Kisii under the Lake Victoria Water and Sanitation initiative. In terms of their depreciation rate these tractors are considered to have a long life span (10 years) as compared to trucks (7 years) and they are also cost-effective given the short haul distance to the disposal site. This means that with proper maintenance and operation, the councils will be able each to handle their waste collection 'in-house' and there would not be significant economies of scale when considering common collection arrangements. Though the councils expressed their desire to manage their own waste collection (to the council officials from interviews conducted, this was a mark of autonomy for each council), without the input from UN-habitat there would be greater economies of scale in inter-municipal cooperation. This option could entail for instance contracting collection out to one firm that would serve the three towns. Another option, purchase of standard vehicles was one of the issues identified for cooperation by the participants at the Kisumu workshop. This is because acquiring just one small tipping truck (an Isuzu NQR for example with the same load capacity as the tractor/trailer) would cost USD 40,000 and above (Manus, 2005), a figure which far exceeds the total expenditure for waste in the municipality of Migori for instance.

Collection by all the three councils is done at no fee at all, the refuse collection charge (RCC) collected from the business and commercial areas are used to subsidise the service to low income areas which cannot afford to pay the collection charges. This, according to field interviews, is considered a fair practice on the basis that the low-income areas purchase their food and clothing from local traders who in turn purchase from the larger traders in the business areas, thus any RCC paid by the commercial areas 'trickles down' to the lower income areas as a cost increase on their purchases. On the other hand though, interviews with the council officials in all the towns revealed that these RCCs are not adequate to cater for the SWM budget in total and this is evident is the inconsistencies of waste collection in the residential areas by the council. Introducing user fees for household waste collection for the in-house collection arrangement in each council would make it more sustainable. This way the councils would be able to take care of the operations and maintenance necessary for the equipment given to them by UN-Habitat next to the other collection tasks.

Transfer stations

Of the three councils, only Homabay municipality has transfer stations in the form of four designated transfer points built in concrete and situated in the CBD, main market and residential estates (see Figure 5.3). The town engineer informed that these transfer stations were constructed with help from UN-Habitat. In Migori, residents put the waste at certain dumping points by the roadside which is then picked up by the municipal workers. In Kisii, certain households served by the private collectors use polythene bags and collection is door to door, while those in the CBD use litter bins and the rest of the households dump their waste at certain dumping points by the roadside which is then picked by the council workers. Participants at both Kisumu and Bukoba workshops showed interest in the technology of transfer stations constructed in Homabay. They looked at it as a best practice that could be shared with other councils after they learnt at the workshop that they could actually seek financial help from UN-Habitat for it.

According to Cointreau (2005), transfer facilities have only modest economies of scale – a small-scale transfer station would handle 80-120 tons/day, thus serving up to 240,000 people/daily shift. With Homabay having its own, the councils can each consider having their own transfer stations. This however, would only work if the locations of the current disposal sites are retained. In the event that a common disposal site for the three councils is considered and that the travel

Figure 5.3. A transfer station in Homabay municipal councel.

time to this site exceeds 30 minutes (Bartone, 2000), then the option of common transfer stations would be more attractive.

5.3.5 Disposal

Most of the waste collected is deposited in open dumps (see Figures 5.4 and 5.5) which as mentioned earlier, are relatively near the CBD and residential areas for all the three municipalities. These dumpsites are potential breeding grounds for disease-causing vectors. The town engineer in charge of SWM in Homabay indicated that the council burns the waste at the dumpsite to reduce its volume but they are still faced with the challenge of the accumulating the ash from the burnt material. Smoke from burning refuse may also be damaging to the health of nearby residents while the smell degrades the quality of life. Of special concern is also in all the three councils, the presence of medical (hospital) waste in the municipal waste stream which poses immense danger especially to the staff working at the dumpsite and those people scavenging the dumpsite.

This situation shows that the current way of disposal is not sustainable. A feasibility study done by CAS consultants in 2005 for Kisii and Homabay already stated that the current disposal sites should face closure in less than five years. However, NEMA recommends a move towards developing landfills whose costs are very prohibitive for such small councils working individually.

Figure 5.4. Disposal site in Homabay municipal council.

Figure 5.5. Designated disposal site in Kisii municipal council.

Based on the studies mentioned, the estimated costs of a sanitary landfill in Kisii municipality for instance, excluding the costs for constructing the access road and acquiring the land, are shown in Table 5.5. These costs include: ground works (excavation and fill), tracks and building (site office/workshop/store), fence, bottom water proofing, leakage drainage system, leakage treatment system and gas collection. The figures show estimated costs for landfills occupying different surfaces assuming the piling depth is 10 m. These costs far exceed even the LATF allocation for Kisii (for instance, in the FY 2008/09 the total allocation was USD 409,573) yet the LATF allocation is not only intended for SWM.

Table 5.5. The estimated costs of constructing a sanitary landfill in Kisii Municipality (adapted from feasibility study by CAS consultants for the Republic of Kenya, 2005).

Surface	Lifespan	Cost in USD
2 ha	5 years	811,800
5 ha	13 years	2,029,500
10 ha	8 years	1,217,700

There are significant economies of scale when it comes to the disposal and treatment of waste. For a sanitary landfill, the economies-of-scale are based on the need to fully utilise heavy landfill equipment that has an adequate compaction ability, as well as the ability to push, spread, grade and cover waste. Typically, a landfill should handle at least 300 tons/day (from up to 600,000 people/ daily shift) (see Cointreau, 2005). These figures can only be realised when the three councils under study cooperate and include the smaller town councils between them.

A common landfill was also one of the suggestions for cooperation raised by the participants at the Kisumu workshop. Therefore considering the amounts of waste, the population served and the costs of construction, a common landfill would be more attractive as the costs can be shared amongst the councils. Issues of location of such landfill, odour and traffic to the landfill will need to be taken to consideration as costs and benefits cannot be distributed equally among all the councils. To come to an agreement, the councils will have to consider tradeoffs. Incorporation of smaller town councils that lie in the midst of these three, who are likely to face externalities from the cooperation, could serve to increase the amounts of waste generated and population served to realise the needed economies of scale. Voluntary agreements would offer a suitable mode to seek their incorporation.

5.3.6 Recovery and re-use

Plastic waste

Recovery and re-use of waste is done on individual (person) basis. In Kisii for instance, there are business men involved in collecting scrap and plastics from scavengers which is then sold in large quantities to bigger firms and transported to Nairobi. Table 5.6 gives details on the quantity of plastic collected per day and the quantity sold from two firms. From interviews with the two firms' representatives, the average quantity sold is calculated taking into consideration weekends and market days when the amount collected is higher than the recorded 300 kgs per day for instance for Ngab scrap dealers and the quantity is usually lower on certain days of the week thus the average figure of 2,000 kgs per week.

There was no evidence of a junk shop or plastic waste collection yard in Homabay and Migori. The officers in charge of SWM in these two councils reported that plastic recycling is done by individuals on a smaller scale and sold at the market on a retail basis for storing paraffin and water.

Though participants at the Kisumu workshop were enthusiastic about a joint recycling plant and getting a viable market for the products, the total amount generated, given that the two firms in Kisii are the major plastic waste dealers, is not enough to warrant a common plant for the three

Table 5.6. Plastic recycling in Kisii municipality.

Recycling individual	Quantity collected	Quantity sold	Cost per kilo
Ngab scrap dealer	300 kg per day	average 2,000 kgs per week	buying at Kshs 8, selling at Kshs 12
Masosa Jua kali	250-500 kg per day	average 3000 kgs per week	buying at Kshs 8, selling at Kshs 12

centres. Instead, the plastic waste from the three towns can be collected and shredded to add value to it and then sold at higher price to processors in Nairobi. This can be done by a private firm.

This conclusion is reached based on reviewing the profiles of firms involved in plastic recycling in Nakuru and Nairobi (Kenya) and in other countries, in order to compare amounts of plastic wastes used and cost of a recycling plant. The profile of a plastic waste recycling firm in Nairobi is given in Box 5.1.

Box 5.1. Green Africa.

This is a plastic recycling firm in Nairobi, Kenya owned by Mr. Evans Githinji. Green Africa recycles plastics into fencing poles; has a staff of 24 employees, and 23 collection yards. They get a supply of 3 to 4 tons daily and per kilo they pay 15 Kenya cents (USD 0.002). It takes fifteen minutes to produce one pole and the firm produces 100 poles per day. The firm owner bought the processing machine which is second hand at USD 14,286. The poles are sold locally and have become favourites particularly to flower farms and Kenya wildlife service because they are resistant to termite attack, to rot caused by moisture retention and to animal attacks. The cost of one plastic pole is Kshs 450 (USD 6.4) compared to a wooden pole which goes for between Kshs 200 to Kshs 250 (USD 2.9 to USD 3.60).

Organic waste

As mentioned earlier, over 70% of the waste generated in these councils is organic. At the stakeholders' workshop, an environmental officer from Migori pointed out that at certain times of the year, there are farmers who request the council workers transporting the waste to dump it on their farms as farm manure. This is done at a small fee agreed between the worker and the farm owner. While this is waste re-use in some way, the fact that it is done unofficially and that the waste is not sorted, poses health risks to the people involved.

According to Cointreau (2005), municipal compost facilities using mechanised equipment would be economic with capacities of 200 tons/day (from up to 400,000 people/daily shift) to enable the full use of loading and turning machinery. Looking at projections on the population and the waste generated in the coming years (see Table 5.4), it would be economically interesting to construct a large compost plant serving the three councils. Considering waste from smaller town councils in the midst of these three would help to augment the amounts and achieve economies of scale.

Research done by Manus Coffey (2005) for UN-Habitat on LVWATSAN shows that the dumpsites in particular those in Kisii and Homabay present possibilities of producing biogas as a recovery measure.

Processing plants such as waste-to-energy plants and anaerobic digesters need to have duplicate process lines so that one keeps on working while the other is down for maintenance and repair. It is noted that for economies of scale, they need typically at least 300 tons/day (from up to 600,000 people/daily shift) (Cointreau, 2005). This again can only be realised when the councils cooperate and put their organic waste together. At the Kisumu workshop the participants learnt about a bio-digester using slaughterhouse waste in Homabay that was put up with the help of UNIDO. It takes up 40 m^3 of waste every week. This ignited discussions on possible options like having the other councils transport their slaughterhouse waste to Homabay or put up a bigger capacity digester or each council having their own digester. The possibility of each council seeking help in terms of resources to construct its own bio-digester like Homabay appeared more attractive to the workshop participants. That Homabay already has its own digester, makes the option of constructing another joint plant unfeasible as Homabay may not join in the venture therefore not achieving economies of scale. Focusing on waste from slaughter houses, it is possible to consider having small size affordable bio digesters that can be put up in each council.

5.3.7 Private sector involvement

The involvement of the private sector in waste collection is minimal and only evident in Kisii. In Kisii, some youth groups are involved in collecting waste from certain residential areas and the council records that these groups have managed to collect 10% of the total amount of waste collected in the town. In Homabay some CBOs take part in street sweeping and the occasional clean-up exercises. The private sector here is however, active in the plastic recycling as discussed above. A novel discovery is the inclusion of the private sector in the recently formed Multi Stakeholder Forums (MSFs) in Kisii and Homabay under the Lake Victoria Water and Sanitation Initiative. The MSFs are made up of representatives from the municipal council, CBOs, NGOs, private entities (business) and other stakeholders. The MSFs are introduced to promote accountability and increase public access to information. These MSFs in the councils are some form of accountability networks (Scott, 2000) that ensure that interventions are developed and implemented in a manner that is informed by and responds to the needs of the local stakeholders. The MSFs are also set up to ensure transparency and corruption-free implementation of projects because among them are members selected to be part of projects implementation units which allow the monitoring of projects' progress and this is necessary for inter-municipal cooperation.

5.4 Possible organisational models for inter-municipal cooperation in Kenya

Based on the study findings on solid waste management problems in these towns and the wider institutional settings (legal and administrative situation) in Kenya, several different organisational models are applicable. These options are presented below.

1. *Joint boards/joint production.* The legislation in Kenya allows for the creation of joint boards, in particular, the Local Government Act Cap 265 which guides the functions of all local authorities. PART VII section 104 -106 of the Act makes a provision for the creation of Joint Boards amongst local authorities similar to those presented by Dollery and Johnson (2005) as 'joint production.' This creation of a joint board would make it possible for the councils to put their resources together and get, for instance, into a joint venture of organic waste re-use. When the councils would put their re-usable waste together, the amount increases which would make it viable to consider the option of putting up a local plant.

2. *Commercial sector, in particular private contracting.* The Local Government Act PART IX section 143 empowers local authorities to enter into contracts. They can jointly seek the services of a private entity similar to what EPA (1994) categorises under the commercial sector. This option would allow the councils, for instance, to source a firm to transport waste from the transfer stations to a common landfill.
3. *Regional organisation of councils.* There are existing regional bodies which provide good platforms for inter-municipal cooperation. There is for instance, the Lake Victoria Region Local Authorities Corporation (LVRLAC) which acts as a Trust for those local authorities who are members. There is also the Lake Victoria Basin Commission.

   Such regional organisations could also work as regional councils as classified by EPA (1994) or what Dollery and Johnson (2005) refer to as a regional organisation of councils. The regional councils provide a regional agenda that complements the self-organising SWM mechanism of the individual member local authorities.

   Given that the three councils all belong to LVRLAC, the councils can with the help of LVRLAC source for funding to obtain SWM resources. They can further harmonise any conflicting goals and objectives of waste management by developing binding agreements. As a starting point towards cooperation, using such an existing regional body to facilitate development in solid waste management is preferable, since the connections between local governments have already been established and financial and technical experience is being built through various projects. Such an organisation is likely to have considerable political and financial independence. This autonomy helps the regional organisation to sustain cooperative partnerships among councils and execute projects in an environment free of individual council politics. Councils could go further and set up an advisory board or establish a reporting structure for the organisation that ensures proper oversight.

Even with such legal provisions, there are no clear specifications of how to avoid conflicts or cases of free-riders as relates to inter-municipal cooperation and in particular, cooperation in solid waste management. As much as informal agreements would offer an easier and flexible option, it is important that the terms and conditions for cooperation agreement are committed to writing. A written document helps to ensure that all parties are aware of their respective responsibilities, and it facilitates review and approval of the agreement by the appointed board or the oversight unit. And from the different options presented, the regional organisation of councils seems to offer more possibilities for cooperation as an immediate solution. This is because such an organisation is already in place and would provide the needed structure to kick- start the cooperation process. This would reduce the logistical costs that would have otherwise been used in setting up a joint board or an oversight organisation to oversee cooperation or even supervise a private firm contracted to provide service jointly to the three councils.

5.5 Modernised mixtures and multi-level governance

In tandem with the arguments of the Modernised Mixtures approach, the multi-level governance concept, not only brings a shift in governance but also seeks to address local challenges in a manner that considers the limited resources available to the small urban centres in the lake basin region thereby ensuring flexibility and robustness.

From the field work, it becomes clear that certain funds for service delivery flow from the national level to local authorities, but that the primary responsibility of waste management lies with the local authorities.

The multiplicity in levels of governance comes in because *first*, not all functions will necessarily be handled in-house, that is each council on its own, and some may have to be handled in cooperation (see Table 5.7):

- local policy determination: single council decides how the waste will be collected e.g. outsourced or in-house), and on which days;
- local management: single council handles the administrative aspects of waste management (i.e. human resources);
- local service delivery: single council delivers the service with staff contracted to it, on its terms and conditions.
- in cooperation policy determination: policy decisions such as how the waste will be treated and when, are made by a group of councils;
- in cooperation management: administration is handled by a body like a joint board or an oversight unit in a regional council;
- in cooperation service delivery: staff delivering the service are contracted to a joint board or an oversight unit in a regional council; rather than individual councils.

The field interviews and the stakeholder workshop revealed that the councils would like to remain autonomous and therefore handle the primary functions in policy making. Service delivery and management on the other hand could be done in house or in cooperation. They were however categorical that waste collection is a function they would like to see remain in-house because they believe this is a first mark of autonomy for each council in SWM.

The stakeholders at the workshop were also in agreement that certain capital-intensive programs like putting up a landfill would be best done in cooperation because of the cost implications

Table 5.7. Possible service provision structures.[1]

Function/ activity	Policy determined	Service managed	Service delivered
Option 1	locally	locally	locally
Option 2	locally	locally	in cooperation
Option 3	locally	in cooperation	locally
Option 4	locally	in cooperation	in cooperation
Option 5	in cooperation	locally	in cooperation
Option 6	in cooperation	locally	locally
Option 7	in cooperation	in cooperation	locally
Option 8	in cooperation	in cooperation	in cooperation

[1] Constructed by the author guided by information compiled from reviewing literature.

involved. Therefore from the 8 options given in Table 5.7, Options 1 to 4 would work well as they allow local policy determination and option 8 on cooperation in all areas comes in where capital intensive projects are involved.

Second, there is also the interplay between a private and a state ordering as explained by Picciotto (2008) as he characterises multi-level global governance. Picciotto argues that regulation typically involves a mixture of legal forms, both public and private, and an interplay between state and private ordering, or, frequently, the emergence of norms with a hybrid status. For instance, for all the 8 options listed in the table above, the councils may decide to involve private contractors in the collection and transportation of waste or even the construction of a landfill. In Kisii for instance, there are already private collectors whose role in waste management could be strengthened by legitimising them and providing assistance like secondary collection of their waste. The existence of multi-stakeholder forums at the council level serves to ensure input from public and private actors.

Local realities, like institutional organisation, resource availability, the presence of non-state actors which are key to the MMA, are therefore also taken into consideration in multi-level governance. These local realities should hereby be described through the lenses of subsidiarity and local responsiveness, economies of scale and externalities, participation, access and accountability to local decisions and equity. In the end, the aim is to seek an appropriate mix of institutions, and financial and technical resources.

5.6 Conclusion

Multi-level governance, particularly its type II form is characterised by task-specific (instead of general purpose) jurisdictions, intersecting memberships and a flexible design that is responsive to temporary needs. In this type II form of MLG, jurisdictions may very well span several (territorially or otherwise defined) levels. This allows this study to consider inter-municipal cooperation to address solid waste management amongst Kenya's small urban centres in the Lake Victoria Basin.

Solid waste management within each of the three councils under study remains a daunting task right from generation to disposal and re-use/recycling. According to the field study, these three councils experience somewhat similar solid waste management problems as is evident in their institutional organisation, the amounts of wastes that remain uncollected, inefficient waste transportation, dumpsites that face closure and recycling/re-use efforts that need to revamped. Efforts from non-state actors where present remain limited. The latest information reveals that as of July 2009, UN-Habitat had donated 3 tractors to Homabay and a similar gesture was to be extended to Kisii under the Lake Victoria Water and Sanitation initiative. This is an early indication that certain functions within the waste chain can still be handled in-house, by each council on its own. However, there are other functions that will need inter-municipal cooperation because of, besides other reasons, the economies of scale. Among these functions are the putting up of a common landfill and a central compost plant. The existing institutional and legislative framework in Kenya allows municipalities to develop such cooperative ventures. There are also organisational models that present themselves for adoption in the event of such cooperation, for instance, joint boards and regional councils. The study presents alongside the discussion certain challenges that would need to be addressed for inter-municipal cooperation to be possible. While there are clear obstacles to inter-municipal cooperation, it also presents opportunities for joint problem solving. Without necessarily being consolidated under one unit, the local authorities can manage some of the SWM problems across jurisdictions using the different methods or approaches to inter-municipal cooperation discussed. Multi-level governance mechanism would thus complement the self-organising mechanism of individual local authorities. It should be noted though that given the different governance arrangements in Kenya, Uganda and Tanzania, these conclusions are specific to the small urban councils in Kenya. These conclusions are particularly applicable to towns that have geographical proximity and share somewhat similar SWM problems.

Chapter 6.
Regional organisations and networks in cross border SWM at municipal level

6.1 Introduction

When local governments confront common regional problems, they can pursue the benefits of coordination with other local entities through regional governance organisations such as regional councils of governments or regional partnerships (Kwon, 2007).With environmental concerns having transcended most political boundaries, Erie and Mackenzie, (2007), posit that there is greater need for coordination across these boundaries to solve transportation, environmental and other infrastructure-related challenges. Building on Chapter 5 therefore, this empirical chapter seeks to answer research question four which is: *What is the role of regional organisations and networks in enhancing cross border solid waste management amongst municipalities in the Lake basin?*

To answer this, the chapter first presents the institutional context for cooperation in the Lake Victoria Basin. Guided by conceptual framework 2, presented in Chapter 2, the institutional context at the regional, national and municipal level is discussed as a first step in looking at cross country municipal cooperation. Here emphasis is made on identifying the constraining and enabling factors to such cooperation. This is followed by a section on regional arrangements under which two regional organisations are presented. These are the Lake Victoria Basin Commission and the Lake Victoria Region Local Authorities Cooperation which are considered as representing the formal and informal regional structures respectively and which have been involved in SWM in one way or another in the lake basin. These organisations are assessed for their role in SWM by the municipalities in the lake basin by looking at municipal autonomy in decision making and resource availability. The aim of this study being to find out how cross border cooperation is actually put into practice and in turn point to feasible options for adequate institutional arrangements in the regional context

6.2 Background-institutional context for cooperation

This section gives a background discussion on the institutional context for cooperation at the regional, national and municipal level, mentioning the enabling and constraining factors to cooperation at the three levels.

6.2.1 The East African community context

The East African Community (EAC) context is the regional context, whereby EAC refers to the Republic of Kenya, the Republic of Uganda and the United Republic of Tanzania (and most recently also Rwanda and Burundi) who together as contracting parties signed for its re-establishment under the EAC treaty of 1999. This history of working together sets the ground for regional and sub-regional cooperation that affects the three countries.

Enabling and constraining factors for cooperation

There are legislative and organisational factors that enable cooperation between the three countries bordering Lake Victoria. The EAC treaty of 1999 certainly sets the pace with its general objective being to develop policies and programs aimed at widening and deepening cooperation among the partner states in political, economic, social and cultural fields, research and technology, defence, security and legal and judicial affairs, for their mutual benefit. Articles 87, 104, 111 and 112 are some of the provisions in the treaty that provide the legal back-up for cooperation in joint project financing, and in the free movement of persons, labour and services. Specific mentioning of environment and natural resource management is done in:

- Article 111 section 2
  a. to preserve, protect and enhance the quality of the environment;
  b. to contribute towards the sustainability of the environment;
- Article 112 section 1
  a. to develop a common environmental management policy that would sustain the eco-systems of the partner states, prevent, arrest and reverse the effects of environmental degradation;
  d. to take measures to control trans-boundary air, land and water pollution arising from developmental activities.

All these provisions though general on trade, movement of persons and environmental issues, cover waste management in one way or another and have a bearing on Lake Victoria Basin which is a shared resource amongst the three (but also including Rwanda and Burundi) as shown in Table 6.1.

Each partner state according to the treaty is to designate a ministry, with ministers who will belong to a council (one of the institutions under the EAC), that acts as the policy organ of the EAC. Today each state has a specific ministry to effect EA regional cooperation. The EAC organs, institutions and laws take precedence over similar national ones on matters pertaining to the

Table 6.1. Lake Victoria surface area, shoreline and basin area per country (FAO, 2000).

Country	Lake surface area		Shoreline		Tributary basin	
	km^2	%	km	%	km^2	%
Kenya	4,113	6	550	17	38,913	21.5
Tanzania	33,756	49	1,150	33	79,570	44.0
Uganda	31,001	45	1,750	50	28,857	15.9
Rwanda	0	0	0	0	20,550	11.4
Burundi	0	0	0	0	13,060	7.2
Total	68,870	100	3,450	100	180,950	100

implementation of the EA Treaty. Emanating from the EAC, are regional organisations that offer a platform for cooperation.

Efforts of cooperation at this level however, are likely to face certain constraints, for instance, in its Articles 111-113, the treaty calls for common laws and strategies on the movement, trafficking and trade in hazardous waste. From the field research in both the primary urban centres and even the small urban centres introduced in Chapter 5, there is no segregation of waste at source or at the point of disposal (except for Jinja now) meaning that the chances of hazardous waste getting mixed with other waste at the disposal site are very high which could affect initial efforts of cross-country cooperation in SWM. This could be the case where for instance, neighbouring councils develop a common landfill and where the public is opposing the setting of such a facility. Yet it is important that the planning and decisions made at this regional level are in line with the practical situation at the local level.

Generally though, it is clear that the institutional context at the regional level already provides a framework that enables cooperation of the three countries under EAC.

6.2.2 National context

The national context like the regional level has constraining and enabling factors regarding cross-border cooperation. The characteristics of the national administrative system (its federal or unitary character, the number of administrative levels and the distribution of responsibilities between the different tiers) and the position of the local government as a second, third or fourth tier in the administrative system of the respective countries provide the point of departure for cross-country inter-municipal cooperation, as they set the conditions that determine the necessity or possible advantages for cooperation in solid waste management.

Enabling and constraining factors

Apart from the regional framework in place, the three countries are also all unitary states with decentralised local governments. As mentioned in the previous chapters, they only have central and local level of government. Within the local levels of government, Tanzania and Uganda have higher and lower ones. Regional units within countries like Tanzania[16] exist for administrative reasons but are not considered sub-ordinate government levels to the state as is the case with the local governments. These similarities in state structures are part of the conditions that make efforts in cross-country cooperation possible.

The level of decentralisation however differs between the three countries as alluded to in preceding chapters. Such differences in the levels of decentralization means that municipalities in the respective countries have different levels of administrative, fiscal and political autonomy. As already mentioned in Chapter 3, the different countries have different conditions related to the

---

[16] In Tanzania, small Regional Secretariats were established (with the enactment of Regional Administration Act No. 19 of 1997) to take the place of the regional development directorates which tended to duplicate the functions and responsibilities of the local government authorities. The Regional Secretariats have been given a redefined back-stopping role to the local government authorities within their area of jurisdiction.

autonomy of budget-making accorded to local authorities by the central government, different systems of staff deployment and different percentages of funds devolved to local authorities. Still, in all the three countries urban authorities are responsible for SWM.

Generally, with regard to the institutional context at the national level, the factors constraining cooperation stand out more because the different levels of decentralisation in the three countries affect the urban authorities in particular where it concerns SWM.

6.2.3 Municipal council context

The Lake Victoria Basin has more than 80 urban authorities.[17] These are categorised into:
- town councils and (Divisions in Uganda) which are the lowest level of urban authority;
- municipal councils;
- city councils which is the highest level of urban authority.

The combined number of city councils is as expected relatively small, there are fewer cities and more town/division councils in the lake basin. The size or scale of the towns may influence their involvement in collective activities. There are also other enabling and constraining factors in line with the level of discretion and the degree of autonomy that municipalities have in carrying out their SWM duties.

Enabling and constraining factors

At the moment, each urban council is managing its own waste (in its jurisdiction), and develops its own by-laws. This situation gives the councils a certain leverage in making certain decision concerning cooperation. There are SWM decisions however, like putting up a landfill that go beyond the level of a municipal authorities and require involvement and permission of a national environmental authority like National Environment Management Council (NEMC) in Tanzania and the National Environment Management Authority (NEMA) in Kenya and Uganda. In Kenya for instance, NEMA is the body that issues licenses for landfills, transfer stations, recycling plants and incinerators. This organisation also issues permits on exporting hazardous waste and requires the exporter to get a prior informed consent document from the national authority of the receiving country.

Within their respective countries, local authorities are allowed to cooperate with each other. What the local government legislations do not explicitly mention is whether this cooperation can be extended to be cross country and therefore a common frame of reference for cooperation in SWM at this level is lacking.

Most of these councils especially for Uganda and Tanzania as discussed in Chapter 3, are also heavily dependent on central government funds and therefore priorities for the use of resources may be more inclined to national programs.

---

[17] Given that membership of LVRLAC only totals up to more than 81 local authorities in Kenya, Uganda and Tanzania excluding Rwanda and Burundi.

The institutional context for cooperation at the municipal level is thus very much dependent on the national level which as shown above is dominated with a number of constraining factors.

6.2.4 Methodology

Two organisations were purposively selected for further study because of their involvement in SWM of municipal councils but also because they give a clear representation of a statutory and a voluntary organisational arrangement.

The data used to inform this chapter were collected through a review of relevant documents with particular emphasis on the East African Community Treaty of 1999, the Protocol for Sustainable Development of The Lake Victoria Basin and the Lake Victoria Region Local Authorities Cooperation Strategic Plan 2009-2014. Two workshops also informed the discussion in this chapter:

- One held in Kisumu, January 2009 that brought together stakeholders from the small towns in Kenya (Kisii, Migori and Homabay) as mentioned in Chapter 5.
- Another in Bukoba in February 2009. It was spearheaded by LVRLAC and the participants came from local authorities in Uganda, Kenya and Tanzania.

Interviews were held with resource persons in the municipalities and the relevant institutions as listed in Chapter 5. Observations on the status of the waste management infrastructure also informed this work. The interviews entailed questions (see Annex 6) that asked about the specific role of regional organisations in waste management (if any) and the existing networks as well as about future possibilities for cooperation in service provision.

Following prior qualitative work from document analysis and information drawn from the administered interviews that informed Chapter 5, together with information from the workshops, this chapter uses an open coding system to codify data. The data were then interpreted and conceptually ordered according to discussions of modernised mixture approach and regional governance. Conclusions drawn on the basis of this research help to bring to light the contributions from this study not only to waste management in East Africa but also to the body of knowledge on regional governance.

6.3 Institutional arrangements for cross country cooperation

Institutional arrangements at the regional context just like those discussed in preceding chapters in municipalities within specific countries, can be either formal or informal. Looking at these different arrangements is important because it is now clear that performance of institution-based development is likely to be undermined or enhanced by the nature of the institutional arrangement.

Figure 6.1 below shows a continuum of regional arrangements showing increasing formalisation from one end with voluntary arrangements which in most cases are informal to statutory arrangements at the other end which are formal in nature. Different kinds of arrangements lie in between these extremes.

The voluntary regional arrangements are such that members (local governments) participate at will and must approve the regional council's activities. The organisations have limited authority to force members to do what they do not want to do. Thus even though a formal institution has been

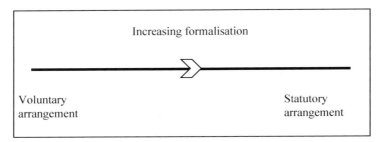

Figure 6.1. Continuum of regional arrangements.

created, its operation remains heavily reliant on self-organising. The specific policy actions that such regional councils take are the product of bargaining and the available mechanisms of collective choice (Gerber and Gibson, 2005 in Sung, 2007). A challenge for this kind of arrangement is the negotiation about an equitable distribution of its benefits which will be affected by asymmetries in economic and political strengths between the actors/members.

Statutory regional arrangements such as, for instance, the regional districts are seen as formalised institutions aimed at promoting regional cooperation. They are likely to impede more ad-hoc forms of cooperation between the localities (see Mullin, 2007). They have less autonomy retained by the individual local government actors involved, in other words, there is reduced local control due to more regulated practices. Where transaction cost barriers to institutional collective action are substantial, voluntary regional governance arrangements may not be possible and statutory approaches such as the regional districts may be more efficacious (Feiock, 2007). Yet again the benefits of voluntary regionalism exceeds the transaction costs when the cooperation produces repeated interactions among the participants, compatible incentive structures, mechanisms to establish reputations, and linkages across various policies and issues.

To assess these arrangements, the following two benchmarks built in the modernised mixtures criteria are used:

- Municipal autonomy in decision making allows the study to establish the flexibility of the regional arrangement to incorporate decisions by municipal authorities in SWM interventions. Inherent in the autonomy is also the power of municipal authorities to act and act collectively, or what Davoudi and Evans (2005) refer to as political capital.
- Resource availability; for a regional arrangement to be viable in the long term, it must have or be able to mobilise resources commensurate with its policy agenda (Davoudi and Evans, 2005). This assures the functioning of an organisation in performing its role in SWM at the municipal level and in turn guarantees its institutional sustainability to strengthen the networks amongst member municipalities. Resources here refer to both material and non-material.

6.4 Regional arrangements on SWM in the Lake Basin

Although there are a number of constraining factors as presented in the preceding section, the study learnt that there are nevertheless several regional arrangements that have been developed so far in the lake basin. These arrangements range from very light structures of mutual information exchange via networking (equivalent of meeting, discussing, starting to coordinate actions in workshops, seminars) to more formal bonding /organisation on SWM. The arrangements involve small urban centres and in some cases a mix of the larger and small urban centres indicating that the variations in scale/size of the urban centres has not been particularly a constraining factor. The different kinds of cooperative arrangement depict some kind of networked polity, where state institutions are working with non-state institutions especially under the voluntary arrangement model of cooperation (Annex 7 shows part of the outcomes of the Bukoba workshop where participants drew a list of the potential and ongoing activities under SWM carried out in cooperation). The study however elaborates further on two regional organisations which represent the opposite ends of the continuum (in Figure 6.1). These two are the Lake Victoria Basin Commission (LVBC) and the Lake Victoria Region Local Authorities Cooperation (LVRLAC) which are categorised by the study as statutory and voluntary respectively.

Under each organisation, the place of the municipal authority within the organisation is discussed. Using the benchmarks of municipal autonomy and resource availability, the organisations are assessed for their role in SWM and finally a typical case of their interventions in SWM is presented.

6.4.1 Lake Victoria Basin Commission (statutory)

The Lake Victoria Basin Commission (LVBC) is a specialised institution of the EAC that is responsible for coordinating the sustainable development agenda of the Lake Victoria Basin. The study categorises it as a statutory body because it was established by the Protocol For Sustainable Development of the Lake Victoria Basin under its Article 33 in 2001 as a permanent apex institution of the community responsible for the lake basin. This means it is legally mandated to perform the activities it is accorded.

The objectives and broad functions of the secretariat of the commission is to promote, coordinate and facilitate development initiatives within the Lake Victoria Basin. The commission envisages a broad partnership of the local communities around the lake, the East African Community and its partner states as well as the development partners.

The place of municipalities within the organisation

LVBC's mandate is spread over three partner states (Kenya, Uganda and Tanzania) and therefore *all urban authorities* in the Lake Victoria Basin are covered by LVBC's work. Recently, Rwanda and Burundi have become part of LVBC too. There is no membership arrangement.

Organisational arrangement

The organogram in Figure 6.2 depicts the organisation's arrangement however, the municipalities do not feature anywhere. This is despite the fact that the commission has a number of programs at the local level. There is a Sectoral Council which is the principal policy and decision making organ for the commission but its members are Ministers from the partner states, which therefore means that the emphasis for the use of resource is likely to be on national programs and not on specific municipal council needs. Municipal councils may therefore merely benefit from programs that have been identified by others at the national level.

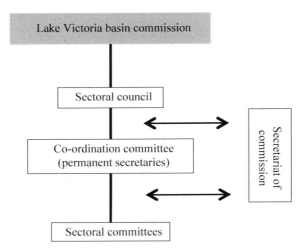

Figure 6.2. Organogram of Lake Victoria Basin Commission (author's construction from field data).

The coordination Committee comprises of all Permanent Secretaries from the three Partner States whose Ministries' mandates relate to the Lake Victoria Basin Commission, particularly Water, Agriculture, Transport, Communication, Energy, Tourism and Wildlife, Fisheries, Environment and Economic development. Again here, the permanent secretaries are the non-political civil service heads in their respective ministries and are thus more aligned to the national level policies and interests. The committee submits reports and recommendations to the Sectoral Council on the implementation of the Protocol for Sustainable Development of the basin.

The Sectoral Committees are composed of senior officials of partner states, heads of public institutions, representatives of regional institutions, representatives of business and industry and civil society. These committees are responsible for coordinating regional activities, preparing comprehensive implementation of programs and submit from time to time reports and recommendations from the working groups. While civil society could have served as a close representation of the local level interests and concerns, field work findings[18] revealed however, that there is not yet a formal collaboration between the Lake Victoria Basin Commission (LVBC) and the Civil Society Organisations (CSOs) who are supposedly part of the committees. There has been no deliberate involvement of CSOs in Sectoral Committees as proposed nor in the working groups and other operational/advisory structures set up to tap into their views on planned policies and strategies.

The secretariat of the Commission whose headquarters is based in Kisumu (Kenya) is responsible for coordinating the preparation, negotiation and implementation of national and regional programs. Specific functions include establishing a regional database and promoting the sharing of information, facilitating research and studies on sustainable development of the basin, mobilising resources for implementation of projects and programs among other administrative duties.

---

[18] Report of the national consultative meeting held between Uganda civil society organizations and LVBC on 30th April 2009, Kampala Uganda

This arrangement on the whole has no place for municipal/local authorities and given the involvement of LVBC at the local level in among others, SWM projects/programs, the study concludes that programs that target municipal authorities are generated or decided upon at the national level.

Municipal autonomy in decision making

The organisational arrangement has a National Focal Point (NFP) in each partner state which are the main links between the specific program(s) and the partner states. They are also responsible for the coordination and harmonisation of the Lake Victoria Basin activities by the various Ministries, NGOs, special interest groups and other development partners in the partner states. Yet even for these NFPs which essentially constitutes the lowest level of contact, the arrangement is still centralised with national ministries from the partner states playing the lead role. This portrays the limited flexibility when it comes to the involvement of urban authorities in decision making within LVBC. Municipal authorities are thus not in a position to act collectively or push for an agenda at a regional level under this organisation's framework and thus infrastructure-related interventions by LVBC are likely to remain restricted to individual municipal councils only. Where the commission has had a local level project like Lake Victoria Water and Sanitation Program (LVWATSAN), they had project management units at the national level, project implementation units at the town level and Multi-Stakeholder Forums. It should be noted though that these units at the local/town level only come in at the implementation stage of a project and not at its initial planning and design stage. The result of this situation is that interventions made at the local level may not necessarily respond to the immediate needs of the local authorities, instead local authorities may just welcome an intervention because it has been made available to them (see Box 6.1 for examples of interventions).

Resource availability and mobilisation within LVBC

Information/knowledge resources

From the field interviews in the small towns, the study learnt that within projects undertaken by LVBC urban councils have benefited from knowledge ranging from technical scientific knowledge presented by experts to experiential knowledge from non-experts shared during workshops.

LVBC has organised the sharing of knowledge amongst a number of local authorities particularly under the Lake Victoria Water and Sanitation program through workshops and exchange visits. LVWATSAN has involved small urban centres. The towns involved all vary significantly in terms of institutional strength and weaknesses and the availability of human and financial resources. Under LVWATSAN, a regional approach to training and capacity building has been introduced so that the towns can learn from each other particularly with regard to case studies, best practices and information on what works and what does not, across the region. Therefore there has been a free flow of information and knowledge amongst participating municipal councils which is essential for collective learning.

Box 6.1. Typical cases of LVBC intervention through LVWATSAN.

Kisii municipality
Kisii is one of the small towns in Kenya that has benefitted from LVWATSAN. Field work in Kisii town revealed that in October 2008 the council was given waste transfer vehicles. A small pickup, manufactured by NDUME Engineering limited of Gilgil, also in Kenya, which uses a two wheeled tractor as a power source and has been modified to suit UN-Habitat's requirements. It has a flat deck body with a very low loading height (see Figure 6.3). The flat body carries eight bins of waste which can be lifted onto or off the pick-up by hand. These little pickups are to be used to collect waste bins from businesses and residential premises and transfer them into large containers or low loading height trailers, which will then be transported to the disposal site. At the time of data collection - Dec 2008-April 2009 - the vehicles were still packed at the council parking area because they had no number plates (official registration) yet and therefore could not be put on the road to use.

Homabay municipality
Homabay is another of the small towns found in Kenya that has benefitted from LVWATSAN. Apart from two pick-ups and collection bins like those received in Kisii town, Homabay also has been donated four concrete waste transfer stations (bunkers) built in the town (see Figure 5.3). As already mentioned in Chapter 5, they are located:
one in the market /CBD emptied daily;
one in Sofia market away from CBD emptied weekly;
one in a site service scheme (residential) and another in Makongeni residential. These are emptied once a fortnight).

LVBC has also organised a transfer of technical knowledge. Under LVWATSAN, for instance, Kisii Municipality in Kenya received small pick-ups for waste transfer and the study learnt that the council drivers benefited from technical training on how to use the vehicles.

Material resources

Finances: The primary funding mechanisms for the LVBC are contributions from the partner states. Funding also comes from development partners. Taking for instance, the budget for financial year 2007-2008 which was USD 2,858,519, partner states contribution was USD 1,664,019 and the funding from the development partners totaled USD 1,194,500. With its funding assured from partner states and development partners, LVBC portrays a more secure financial standing even as relates to its role in SWM than a voluntary organisations whose funds are dependent on membership subscription.

Staff: The functioning of the national focal points (NFP) in each partner state requires full-time staff. From the field work, it was evident that under the current structure, the NFP officers in addition to the Lake Victoria Basin Commission have also other national responsibilities and

Figure 6.3. Pick-up for waste transfer.

hence are seriously overstretched and accordingly need beefing up. On the other hand, these staff are paid by partner states relieving LVBC off this burden and thus allowing LVBC to use its resources to undertake other things, including capital projects under SWM as is exemplified in Box 6.1. LVBC as an organisation is able to develop a ring fenced budget specifically for SWM.

Therefore from the angle of resource availability, material and non-material, LVBC is well positioned to intervene in SWM as a regional organisation but its interventions so far have remained oriented towards individual councils and not on cross-border cooperation.

These interventions bring out the supporting role of LVBC. It is actually providing back-up to the small urban centres in carrying out their traditional SWM tasks of collection, transportation and disposal of solid waste. LVBC has provided transportation vehicles and even built transfer points. The true success of these interventions in Box 6.1 can only be measured in the long term. The study however, questions the interventions as to whether they actually responded to the immediate needs of the municipal councils or whether the municipalities just welcomed the interventions because they were made available to them. This question is raised because taking the case of Kisii Municipality for instance, the vehicles were lying in the packing area more than six months after delivery which points to the lack of urgency for this intervention. Further, according to the Manus Coffey 2005 Consultancy report on SWM in Homabay and Kisii municipalities, organic wastes which remain in the corners of the bunkers decomposes thereby causing odour problems and also harbour a reserve of bacteria which then speed up the start of decomposition in subsequent wastes. All wastes must be removed completely at least every two days if the bunker is not to become a source of health problems. It should be noted though that the bunkers in the residential areas are emptied only once a fortnight and that even the bunker in Sofia market is only emptied once a week. See Figure 6.4 of an overflowing bunker in a residential area. This points to the low level of ecological and institutional sustainability of the interventions made.

Generally though, LVBC being a formal institution has had a number of strengths particularly when it comes to its access to resources. LVBCs sources are assured. As a result LVBC is able to develop interventions even in large scale SWM infrastructure thereby playing a supporting role to the small urban centres. That there is no place for the municipalities in this organisational arrangement however, points to the limited autonomy of the municipalities to take part in decision-making and in turn influence the kind of interventions they receive. It also depicts their inability to act collectively with other municipalities under the LVBC framework.

Though the interventions so far have been within individual municipalities only, there are clear indications that this formal nature of LVBC and the supporting role it plays so far is likely to influence the role it would play when enhancing cross-border cooperation in infrastructure.

Figure 6.4. A bunker overflowing with waste.

6.4.2 Lake Victoria region local authorities cooperation (voluntary)

Lake Victoria Region Local Authorities Cooperation (LVRLAC) was formed in 1997 as a regional cooperation of Local Authorities in the Lake Victoria Region on the initiative of the mayors of Entebbe (Uganda), Mwanza (Tanzania) and Kisumu (Kenya). LVRLAC's main role is to facilitate local authorities in responding to concerns of sustainable management around Lake Victoria of which waste management is one. It is categorised as a voluntary arrangement in this study because membership is open and authorities are free to join as they deem fit. Actually a main attribute of voluntary regimes is their flexibility and because of this characteristic, LVRLAC has developed into a growing network organisation with over 81 local authorities from the Lake Victoria Basin of East Africa - in Kenya, Tanzania and Uganda. Contacts have been made with the governments of Burundi and Rwanda, and the intention is to also incorporate local authorities from these countries in the future.

With the focus being on urban authorities, members range from very big size authorities with populations of over 500,000 people, for instance, Mwanza city council in Tanzania, Kisumu city council in Kenya and Kampala city council in Uganda to the small town councils with a population of less than 100,000 people, for instance, Ahero town council in Kenya as shown in Table 6.2. Town/Divisions councils are the majority in terms of LVRLAC's urban authorities membership, probably because there are many more townships compared to cities and municipalities in the lake basin. The diversity of membership is a key strength for LVRLAC as this can be a rich source for knowledge and information exchange.

The place of municipalities within the organisation

The network governance structure of the organisation as shown in the organogram in Figure 6.5 has a General Assembly (GA) constituted by all subscribing members (the majority of whom are the local authorities) as the supreme decision making organ of the organisation. The GA is

Table 6.2. Local authorities in numbers and according to levels.

Level/scale	Kenya	Uganda	Tanzania
City council	1 (Kisumu)	1 (Kampala)	1 (Mwanza)
Municipal council	6 (e.g. Homabay, Kisii, Migori, Siaya, Busia)	5 (e.g. Jinja, Masaka, Entebbe)	3 (Musoma, Bukoba, Shinyanga
Town/division Council	7 (e.g. Ahero; Awendo; Rongo; Oyugis; Bondo)	27 (e.g. Lukaya; Busia; Mukono; Mpigi;Rakai; Kira; Lyantonde)	-
District council (rural)	-	4 (Kalangala, Mayuge, Mpigi, Wakiso)	20 (e.g. Bariadi, Bukoba, Bunda, Chato, Gheita)
County rouncil (rural)	11(e.g. Nyando, Suba; Rachuonyo)	6 (e.g. Buwama, Kasanje, Katabi)	-

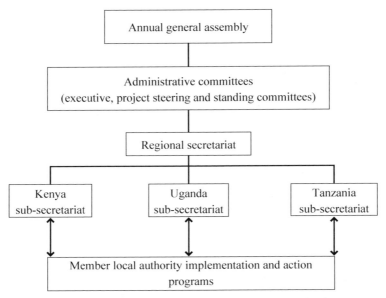

Figure 6.5. Organogram of LVRLAC (author's construction from field data).

responsible for determining the overall direction of the network. Compared to the LVBC, this arrangement presents a more decentralised organisational model which affords the local authorities a key position in the organisation. The GA elects an Executive Committee which is the policy making organ of the network with the responsibility to oversee the day-to-day activities of the organisation. There are other administrative committees that are part of the organisation as well (including project steering and standing committees). The management function of the network is vested within a regional secretariat currently based in Entebbe, Uganda and supported by three country-based sub-secretariats in Kampala (Uganda), Kisumu (Kenya) and Mwanza (Tanzania). In essence LVRLAC has a regional framework under which country offices are linked.

Municipal autonomy in decision making

Being the constituting members of the annual General Assembly which is the top organ within the organisation, then certainly the municipal authorities have more autonomy in decision making as compared to their position in LVBC. These members have statutory/legislative and administrative mechanisms at their disposal which provides great potential to enlist policy changes at the local level within the Lake Victoria Region with LVRLAC playing an enabling role. Yet there are always differences to contend with. Despite their numerical dominance at the top level of decision making and even a shared policy agenda on promoting sustainable development in the Lake Victoria Basin, political, administrative and fiscal differences amongst members influence their capacity to act collectively. Where acting in this case would mean deciding and implementing a regional waste treatment plant for instance, so far the interventions under a common regional framework related to SWM infrastructure have been carried out in individual urban authorities and not collectively.

Resource availability and mobilisation within LVRLAC

Information/knowledge resource

Like LVBC, members of LVRLAC under specific projects, have benefitted from both technical and experiential knowledge. Member councils involved in the pedagogic program (which is described later in this chapter) for instance, have received training on how to set up and run the pedagogic centres.

A number of workshops including the Bukoba that informed this study have allowed sharing and exchange of information. The Bukoba workshop for instance was on regionalising waste management and enabled the participating member councils to share their experiences on best practices. See Box 6.2 on a part of the outcome of the Bukoba workshop.

While the organisation is committed to improve information and knowledge sharing (as put down in its strategic framework) among its members, not all members under LVRLAC have benefitted from the information sharing. Just seven member councils took part, for instance in the pedagogic program whose selection for inclusion according to an interview with project officer, included amongst other things resource endowment of a particular council and having an active department that addresses SWM.

The study further learnt that despite the existence of a monitoring, evaluation and reporting (MER) system, there is limited practical application of the system due to poor data collection, management and sharing between the LVRLAC secretariats and the member local authorities most of whom do not have the modern ICT-equipment. This makes effective network performance measurements difficult as it limits to a great extent the possibilities for feed-back from and continuous dialogue between the organisation and its members.

Box 6.2. Information sharing: Bukoba workshop in February 2009.

Exemplary practices on SWM amongst urban authorities in the Lake Victoria Basin:
- Community partnerships, micro/small-scale enterprises in solid waste management (sorting at source, composting, recycling), community contracting, Municipal support mechanisms, e.g. Kisumu Integrated Solid Waste and Management Programme (KISWAMP).
- Use of technical methods (for example composting, ecological sanitation – ECOSAN and biogas), e.g. Kampala City Council ECOSAN and Homabay Bio-digesters.
- Public-private-community partnerships, e.g. Mwanza City Council contracting.
- Commercialising of waste collection and transportation mechanisms, e.g. Kampala City Council landfill management and private waste collection (skipless system).
- Innovative disposal methods (composting, biogas, semi-engineered and sanitary landfills), e.g. Kampala City Council.
- Strengthening financial and economic implications of solid waste management, e.g. KISWAMP.
- Municipal capacity development and sensitisation on waste management, e.g. CDS (financing toolkit).

What could be beneficial however, in the LVRLAC arrangement is the diversity of its membership, shown in the different sizes of the urban centres, which at any one seating can become a rich source of information/knowledge. The Kisumu workshop for instance allowed participants from smaller municipal councils to learn from Kisumu City council's experience under the Kisumu Integrated Solid Waste Management Plan specifically on the component of 'waste is money'. These small councils were encouraged to seek opportunities to train small voluntary groups to look at SWM as an opportunity to make money.

Material resources

Finance: the members which are local authorities pay a membership fee and an annual subscription as shown below:
- full membership fee: USD 1,500 (paid once);
- annual subscriptions:
  - district/city councils: USD 500;
  - municipal councils: USD 400;
  - town councils/divisions: USD 300.

There is also the possibility for associate membership of LVRLAC open to any cooperation, authority, organisation and individual that is interested in promoting the use of Lake Victoria resources for public benefit, but does not meet conditions for full membership.

The project officer informed the study that the organisation cannot however, rely on the members subscriptions because it is too little. The organisation has an annual budget of general operation that comes to an average of USD one million and the organisation therefore has to raise funds from donors like SIDA.

Staff: LVRLAC maintains a small staff structure and has been working on the assumption that the paid staff of the member local authorities will backstop the work of the network without any extra remuneration (voluntarily). LVRLAC's project officer informed the study that this however, is not the case. The study learnt that, even the identity, roles and image of sub-secretariats hosted within the member local authorities are still unclear and poorly understood by the member local authorities. There have been experienced delays in the provision of office space and services by those members hosting parts of LVRLAC administration. For instance, when the operational procedure of the sub-secretariats are not compatible with the bureaucratic processes of the host member local authorities. Some of the Network organs specifically the standing committees are reportedly dysfunctional.

Generally, resources both material and non-material under LVRLAC appear limited. This has had an impact on its role in SWM within individual urban authorities. From field work it became clear that certain projects undertaken have stalled and this is seen in one of the cases under box 6.3.

These interventions point out to the supplementary role of LVRLAC. Compared to LVBC, LVRLAC's role is more one of adding on to the traditional tasks of SWM by the local authorities. These interventions are essentially demonstration centres on what are considered best practices on SWM like waste re-use. The interventions in smaller towns (Masaka, Homabay and Musoma) presented in Box 6.3 show successful connections with the departments in the councils dealing

---

Box 6.3. Typical LVRLAC's intervention through environment pedagogic centres.[1]

Kisumu City Council in Kenya
Kisumu was the first environmental pedagogic centre (EPC) to be established and became fully operational in 2004. It reached the most advanced stage as EPC but momentum of the project slowly dwindled over time and at the time of field work (2007/08) there was nothing much to show for it on the ground. The reasons for this failure varied but they include inadequate human capacity, budget restraints because a lot of money went into paying external consulting agencies and also the political situation where there was violence and destruction of property after elections.

Masaka municipal council in Uganda.
The EPC was initiated in May 2008 and its status is 85% complete. EPC's new construction is adjacent to the Public Health Department building in the town centre hence highly accessible. Hosted by the environment department, the facility is fully integrated into the functioning of the department and will be applied to demonstrate best practices in environmental management. The EPC constituted demonstration garden plots, a composting unit, a solar powered main hall, an exhibitions hall, office space, a resource room and roof water harvesting.

Homabay municipal council in Kenya.
The EPC was initiated in March 2007 and its status is completed. Homabay's EPC is very suitably located within the town centre at a site earmarked for future expansion of the council offices. It has enrolled participation of a youth group that also uses the centre as learning ground on various innovations and environmental practices. The tree nursery supplies the council with seedlings for its town greening program while schools often use the facility as a practical learning centre. The centre is constituted of demonstration garden plots, a greenhouse, a tree nursery, water harvesting facilities, composting, a solar powered building and a host of information displays including one on bio-digesters.

Musoma municipal council in Tanzania
The EPC was initiated in June 2008 and its status is 90% completed. It is a unique EPC in the sense that it uses the concept of a recreational learning centre by establishing different components of the EPC within a park environment. It is strategically located within 200 meters of the town hall and though not fully completed yet, it has fully established a vibrant tree nursery that is supporting the town greening initiative. It has also attracted partnerships with CBOs dealing with environmental management.

---

[1] The EPCs are part of the Swedish support to the Lake Victoria Initiative (LVI). Centres are established with the aim of demonstrating to the population how various environmental issues can be dealt with, for example through general information on environmental conditions and challenges.

with environmental issues and an integration with ongoing plans and projects. It remains to be seen whether these centres will remain active in the long term or whether like the one in Kisumu, they will stall.

Overall, LVRLAC as an informal institution has allowed municipal authorities a key place in its organisation arrangement and in turn in its decision making. The informal nature of LVRLAC is associated with the massive political will of member local authorities. Their numerical dominance notwithstanding, these municipal authorities have not been able to act collectively to cooperate in infrastructure provisions. Apart from the differences (politically, administrative and fiscal) inherent to the diversity in the member councils, the weak state of its resources does not sustain the supplementary role of LVRLAC in developing SWM interventions or in steering collective action. Its interventions so far have remained implemented in individual municipalities only.

6.5 Conclusion

There are a number of enabling and constraining factors for cross-country municipal cooperation in SWM. These factors are evident in the institutional context at the regional, national and municipal level within the three countries in the Lake Victoria Basin. The enabling factors are the supportive legislation at the regional level as well as existence of regional organisations that could provide a platform for cooperation. Constraining factors are the differences in the levels of decentralisation in the three countries and even different municipal by-laws. These factors directly affect the efforts of cooperation. There is however, still a number of regional arrangements that have been developed so far though most of these are informal or of a voluntary nature and do not involve cooperation in infrastructure provision on SWM. Two regional organisations (LVBC and LVRLAC) were assessed for their role in enhancing cross-country SWM in an attempt to analyse how cross-border cooperation is being put into practice and to conclude on feasible institutional options for SWM at the regional context.

LVBC is categorised by the study as statutory and thus formal. As an organisation, it is looked at as an enabling structure that can provide the platform for cooperation having been formed within the East African Community. However, it does not have a place for municipal authorities in its organisational arrangement meaning that municipalities have little if any influence on decision-making on SWM interventions by the organisation. It also means that municipalities have not had the power to act collectively under LVBC. LVBC however, has both material and non-material resources, which safeguards its supporting role in SWM. Interventions in SWM infrastructure provision so far under LVBC though under a common framework, have been realised in individual municipalities only.

LVRLAC on the other hand is categorised by the study as voluntary and therefore informal. It is an enabling structure to cooperation just like LVBC. The organisation's annual general assembly is made up of local authorities who are its ultimate decision makers. Thus municipalities have a key position under the LVRLAC arrangement. However, even with their numerical dominance, municipalities have not been able to act collectively in infrastructure provision. The study links this occurrence to the inherent differences (political, administrative and fiscal) among the member councils but also to the informal nature of LVRLAC's operation which presents a constraining factor to effective cooperation. In its supplementary role, LVRLAC has lacked sufficient resources, material and non-material, and this does not empower it to steer its members to act more collectively. Interventions in SWM infrastructure provision so far, just like under LVBC have been in individual municipalities only.

The institutional context particularly at the regional level brings out higher expectations for cross-border cooperation in SWM. This is particularly the case because the East African Community and the regional organisations are already in place. In reality, there is however, relatively little cooperation and most of it is in non-material aspects. The presence of this evidence for regional cooperation though in non-material aspects, points nevertheless to the possibility of further cooperation in SWM infrastructure provision in the long term. The two organisations - LVBC and LVRLAC - portraying formal and informal arrangements, can thus be looked upon as playing different roles when it comes to infrastructure provision. Both of them would offer services and infrastructures based on their specific strengths. The formal organisation would doubtless be more satisfactory from a legal standpoint, as it would allow all the participating municipalities to benefit from the legal stability essential to the implementation of joint activities. Given its position with the national governments and partners, the formal organisation could play a key role in getting or mobilising resources for infrastructure provision. Yet, to have a meaningful impact, it would have to reconstitute its organisational structure to bring the local authorities on board and to allow them to take part in decision making. This transformation would serve to ensure that inventions made are in line with local needs. The informal organisation is more flexible and would allow more easily for local self-governing. It would thus play a key role in pulling the municipalities together to act collectively particularly in information and knowledge sharing. But even then it needs to seek innovative ways of getting more financial resources in addition to the membership subscriptions. This way it will provide a stronger platform for cooperation.

Chapter 7.
Conclusion and discussion

7.1 Introduction

This chapter formulates the conclusions when answering the research questions that were formulated at the beginning of this study. The chapter begins by giving an overview of the framework under which the study was built including a brief explanation of the conceptual models, research questions and methodology. Conclusions made for each preceding chapter are then given. Section 7.2 begins by looking at the internal organisation of three municipalities (Kisumu, Jinja and Mwanza) to assess their performance of SWM tasks. This section also looks at their collaboration with non-state actors at municipal level to compare the different service arrangements. Section 7.3 deals with collaboration amongst municipalities by assessing the opportunities for inter-municipal cooperation. It also looks at the role of regional organisations and networks in enhancing cross-country SWM cooperation at the municipal level. Section 7.4 presents some concluding remarks.

7.1.1 Study overview

This thesis aimed at contributing to the improvement of solid waste management at the benefit of the urban poor in the Lake Victoria Basin in East Africa. More specifically, it focused on institutional arrangements for infrastructure and service provision at the municipal level. With the current global trends towards decentralisation of national governments' responsibilities, then certainly the local/municipal level and in turn municipal authorities take centre stage. There is much agreement that municipal authorities have not performed adequately, yet it is also clear that they remain the everyday face of the public sector; the level where essential services are delivered to individuals and where policy meets people (World Bank, 2000). To take it even further, municipal authorities are the legal owners of waste once it is put out for collection (Schubeler, 1996). Taking a bioregional perspective therefore, that is, covering the Lake Victoria Basin, this study sought to analyse SWM arrangements in different municipal authorities in the three countries found in the lake basin in order to gain insights and present feasible options for institutional arrangement for SWM at the municipal level.

Two conceptual frameworks (Figures 2.1 and 2.2) were thus developed to guide the study towards identifying different institutional layouts. At the centre of both conceptual frameworks is the municipal authority.

The first conceptual framework (Figure 2.1) shows the different elements of the SWM tasks under the full or partial responsibility of municipalities. Secondly, it illustrates that a municipal authority when seeking to fulfil the SWM task and responsibilities enters into a number of different relationships with other actors both formal and informal. Thirdly, it shows that municipal authorities operate in between the national and neighbourhood/ward levels wherein the households are found. Fourth and finally, the figure indicates that the actual SWM arrangement at municipal level can be judged with the help of three kinds of criteria: flexibility, accessibility and institutional and ecological sustainability, all of which have been operationalised to reflect the local conditions under study.

The conceptual framework on SWM at municipal level (Figure 2.1) was developed for national and local SWM situations within a particular country but SWM issues could also transcend country boundaries and be looked at as cross-country cooperation. And this is why the study developed a second conceptual framework, shown in Figure 2.2. This second conceptual framework brings on board the regional concept which allows the study to look at cross country municipal cooperation under the umbrella of the East African Community. The framework is broken down to look at enabling and constraining factors for cooperation at the state level, at the municipal level and specifically under SWM infrastructure.

The criteria of flexibility, accessibility, institutional and ecological sustainability are found within the Modernised Mixtures Approach (MMA) which has been used to adapt Ecological Modernisation Theory for the study of SWM and sanitation. Under MMA, SWM systems have been deliberately and reflexively reconstructed in response to the challenge of a changing social, economic, and environmental context in East Africa (Spaargaren *et al.*, 2006). Intelligent combinations of formal and informal, public and private, citizen participation and professional management then become evident. With this understanding, the following research questions were formulated to structure the study:

1. What is the current status of the (physical) environmental infrastructures and the level of service provision for SWM in three designated urban centres in EA namely Kisumu (Kenya); Mwanza (Tanzania) and Jinja (Uganda)?
2. What are the existing policy arrangement for SWM in the three urban centres and what can be learnt from the differences and/or similarities amongst them?
3. What are the possibilities for cooperation in solid waste management amongst small neighbouring municipalities in Kenya namely Kisii, Homabay and Migori municipalities?
4. What is the role of regional organisations and networks in enhancing cross border infrastructure provision in solid waste management amongst municipalities in the lake basin?

Chapters 3 and 4 answered the first and second research questions by looking at three urban authorities in the Lake Victoria Basin: Kisumu City Council in Kenya, Jinja Municipal Council in Uganda and Mwanza City Council in Tanzania. The three were selected for comparison because of similarities in urban status – they are all regarded as primary urban centres as they come just below the capital cities in their respective countries. This in turn means that differences in public service provision levels and population dynamics are considered minimal. The three are also found in the Lake Victoria Basin. These general similarities allowed for a comparative method to learn from the differences and also the similarities existing in their SWM systems. To make an in-depth comparison, a case study research design was employed. The three urban centres thus provided multiple cases bringing out insights which would have otherwise been lost in a single case study. The study was able to discuss exemplary practices as well as shared problems/obstacles in SWM. Building on the findings of the three towns, the study went ahead to look at three other smaller urban centres but this time all three in Kenya. The three served as embedded units within the Lake Victoria Basin and they are Kisii municipal Council; Migori Municipal Council and Homabay Municipal Council. Work done in these three small urban centres informed Chapters

5 and 6 and in turn answered research question 3 and part of research question 4. The role of two regional organisations (Lake Victoria Basin Commission and Lake Victoria Region Local Authorities Cooperation) in enhancing cross-border SWM in the lake basin was also assessed and answered the other part of research question 4.

Within the cases under study, both qualitative and quantitative methods of data collection and analysis were used. There were data that could be drawn from household surveys while other data came from interviews, stakeholder workshops, document analysis and observation. Combining both methods allowed the study to benefit from the flexibility of qualitative methods to do an intensive examination of the cases as well as from the advantage of generalising the findings to other urban authorities in East Africa made possible by the quantitative methods.

7.2 Internal organisations and collaborations with non-state actors

7.2.1 Performance of tasks

In response to research question 1 which is on the current status of (physical) infrastructure and the level of service provision for SWM, the study looked at the internal organisation of the three primary councils mentioned earlier focusing on their performance of tasks under both the technical and social dimensions along the waste chain. The technical dimension refers to waste flows together with the infrastructure and technologies and their direct planning and management. The social dimension captures the relational capacity of municipalities, specifically the qualities of relationships municipal authorities engage in in order to have the tasks under the technical dimension performed well. Therefore issues to do with public/citizen participation, public-private partnership and social trust are key here. Interviews with resource persons, observing the SWM infrastructure and document analysis were particularly useful for this section.

Looking at the performance of the three councils, the study concludes that first the three municipalities 'share' an overall level of performance that is rather low. In the SWM policy cycle, these East African municipalities are still in their infancy phase, what is referred to in the modernisation discourse as a phase of simple modernity (see Scheinberg, 2008, Scheinberg and Mol, 2010). It is not a fully developed management system yet. This is explained in the collection rates which are low to average with Kisumu recording 25-30%, Jinja has 40-60% but Mwanza stands out with 88%. There is no evidence of efforts to organise at source separation/recycle or re-use by the municipal authorities themselves (except, very recent, the case of Jinja in composting). This can be interpreted to mean that achieving ecological sustainability has not been a priority for these municipal authorities. All the three still use large vehicles which are inadequate in number but also to handle the high density kind of waste generated. They all use open dumping grounds and except for Kisumu, waste management is still lumped with medical health issues in the department of public health and it is not seen as a distinct environmental problem.

Nevertheless they have differences in the performance of the tasks with some being centralised, others decentralised and in some cases a mix of the two with different measures of success. This leads to the second conclusion which is that there are no fixed routes/practices in the performance of SWM tasks instead there are lots of variations. This in turn means that there is room for manoeuvre at municipal level to improve on performance of SWM. These improvements are being made in the different towns already. Jinja and Kisumu are cases where the municipal authority has had relative autonomy in decision making to organise SWM accordingly. This is reflected in the involvement of even supra local actors where UN-Habitat and other organisations are working together with Kisumu city council to develop an Integrated Solid Waste Management Plan. In Jinja, the World Bank and NEMA are working with the council to develop a compost plant at the disposal site. This means that the council wants to make an additional step in addition to the traditional tasks of waste collection, transportation and disposal. Mwanza shows how improved access to both material and social resources improves SWM. Though limited in resources (financial and equipment) for SWM, it has with the guidance of its by-laws, made use of social resources. The council therefore only has to do or pay for cleaning of public areas and management of the disposal site in addition to supervising the contractors. This, as also mentioned by Awortwi (2003), worked to reduce the social and financial burden of the municipal authority (where Kisumu had an expenditure of USD 420,000 for SWM in year 2006/2007, Mwanza spent only USD 150,643 in the year 2007). This means that relative autonomy in decision making and access to resources both material and social are some of the possible routes for manoeuvre to improve the performance of SWM on the whole.

Third, municipalities which embrace SWM as a task with only the technical dimension and little consideration of the social dimension, tend to show a lower performance. This is evident in the percentages of waste collected, service coverage and differences in municipal resource needs for SWM in the three urban authorities. Of the three municipalities, Mwanza has incorporated the social dimension and because of that, collection is higher at 88% compared to Kisumu where the municipal council is wholly responsible for SWM with a waste collection percentage of between 25-30%. Jinja comes in the middle with between 40 to 60%. Kisumu council being solely responsible, has to employ more staff and is currently facing a deficit of 34 staff members to make up the optimal number of 71 whereas the staff in Jinja including the casual sweepers are just about 30 in total. In Mwanza there are about 55 workers in total. Kisumu also needs to incorporate more resources (finance and equipment) compared to the other two councils who are benefiting partly from resources of the other public and private actors. Therefore to improve on performance of SWM tasks, municipalities have to look at SWM as a socio-technical system integrating both the technical and social dimensions.

7.2.2 Collaborations (formal and informal)

Research question 2 considers the existing policy arrangement for service provision. And in response to this question in addition to findings under research question 1, the study assessed the formal and informal collaboration with non-state actors. Hereby the focus was on issues of legitimacy and influence on decision making, relations and alliances between the municipality and the non-state actors, amongst the non-state actors themselves and between these actors and the households and also the payment systems. In this section, the debate between the neo-developmental state and network governance was key. Interviews held with the different SWM actors and household surveys were particularly useful. The following conclusions are made:

First, the different forms of collaborations show different levels of success. This is evident in Table 4.12 which shows the percentages of households receiving service, service satisfaction amongst households and the percentage of waste collected, all of which are indicators for the success or failure of the SWM system. This is the case where Mwanza has a formal community dominated arrangement but also with the involvement of some private firms in the CBD and the municipal authority. Jinja has private firms working formally with the municipal authority while in Kisumu, it is the council but with informal involvement of both community and private entities. This supports the arguments of EMT for inclusion of 'modern' institutions in this case, the private firms(markets) and the CBOs (civil society) as we seek solutions to environmental problems.

Yet despite these collaborations, municipal councils are still the loci of authority in SWM at the municipal level. In all the three councils under study, it is the council authorities that make decisions on SWM, award and terminate contracts, monitor/supervise contractors and in some cases make payments to the contractors. The data gathered from the interviews and responses from households indicate that leaving SWM solely to the private sector raises concerns of equity particularly to the urban poor, and leaving it solely to the community raises concerns of sustainable service provision. But from another angle it could be viewed to confirm what Moyo *et al.* (1998) write about the presence of some kind of residual resistance or lack of enthusiasm on the part of local/municipal authorities to share responsibilities with private sector enterprises. Nevertheless, it remains clear that municipal authorities as institutions are still needed to take a lead in SWM at the municipal level.

Third, municipalities that incorporate non-state actors formally, show better performance than those municipalities that ignore this option. And this has been particularly the case where non-state actors entail both the civil society and the private companies because this mix of different groups captures different social and economic realities on the ground. To support this, Table 4.12 shows the field findings where the percentage of households receiving SWM is higher in Mwanza at 82.5% compared to Kisumu at 46.5% where the arrangement is informal. In Jinja where the municipal council has only incorporated private companies, the percentage of households served is 60%. It is easier for the municipal authorities to monitor the performance of formal arrangements because the actors are under their supervision.

Fourth, even with formal incorporation, the duration of the contract arrangement in relation to the type of non-state actor contracted affects performance. The one year contract arrangement is considered short in both Jinja and Mwanza and the effect of this is seen especially in Mwanza where competition amongst contractors is not evident per se. The majority of the contractors in Mwanza are small CBOs with little initial investments trying to establish a niche in SWM. The uncertainty of winning the contract come the following year has resulted in little investment in SWM (see Karanja, 2005 and Awortwi, 2003).

Finally, a balance between legitimacy and efficiency serves to improve SWM. While the legislation and the legality of the procedure of contracting out service provision is a source of legitimacy for the municipal authority and the formal non-state actors in Mwanza and Jinja, public acknowledgement also allows achievement of higher levels of legitimacy. Public acknowledgement comes from efficient service provision which is reflected in this study in the performance of the SWM tasks in terms of percentage of waste collected, service coverage, service satisfaction and even percentage of households that pay for service provision (see Table 4.12). Kisumu shows a case where informal actors though not recognised as legal SWM service providers, are legitimate in the eyes of the public. This is evident from the percentages of service satisfaction (70.6% where n=93).

The balance between legitimacy and efficiency for the case of Kisumu could therefore be achieved by linking the municipal authority with these informal non-state actors. These actors have control over a significant proportion of the population and they have shown (internal) consistency with 71% of those interviewed having been in the solid waste sector for more than five years already.

The findings presented in Table 4.12 show remarkable differences in the three towns. While Mwanza has taken the leading position in terms of percentage of households receiving SWM, percentage of households that pay for waste collection and percentage of waste collected compared to the other two towns, service satisfaction amongst households in Mwanza is lower. The findings do not give definite explanations of this result but picking from the list of recommendations by households on how services can be improved, then a probable explanation could be high levels of expectation from households. It is possible that households are not just content with high collection rates but are keen on the quality of service given as well. The satisfaction rate could also be associated with the duration of the contract which is considered short and thus a deterrent for contractors to make investments to improve on their performance of SWM. The level of professionalism of the CBOs may also be wanting given that a number of them are involved in other activities than SWM to sustain their incomes. But this suggestion remains to be tested in further research.

7.3 Inter-municipal cooperation

The study discussed Inter Municipal Cooperation (IMC) as involving a number of local authorities, or municipalities, in proximity to one another, which join forces to work together on SWM. These municipalities could be in one and the same country which is presented in section 7.3.1 or they could be from different neighbouring countries as presented in section 7.3.2.

7.3.1 Cooperation within a country

Research question 3 seeks to find out the possibilities for cooperation in SWM amongst smaller neighbouring municipalities. To respond to this question, the study assessed opportunities along the waste chain for IMC amongst three small urban centres in Kenya. Data that informed this section were obtained from interviews with resource persons and a stakeholders' workshop in addition to document analysis. There are two conclusions on IMC in SWM in Kenya.

First, IMC offers potential for improving SWM amongst small urban centres. This is because these small urban centres have low tax base and in turn low revenues. The total transfers from central government to the three small councils under study in the financial year 2008/2009 were: Migori - USD 427,158; Homa-bay - USD 456,551; Kisii - USD 409,572 and yet this is not only intended to be used for SWM. Take Migori for instance, whose total expenditures on SWM in the year 2007/2008 were USD 41,966 while the transfer from central government was USD 381,009 that year. The amount allocated to waste management is not enough to purchase even a waste collection tractor at USD 40,000 for instance and still pay workers and have a balance for maintenance and repairs. This in combination with the amounts and types of waste generated, population figures, geographical proximity and minimal differences in the amount of funds transferred from central government transfer to the three councils support the conclusion that

IMC would offer potential for improving SWM specifically regarding waste transportation vehicles, waste disposal infrastructure and waste treatment facilities as they all present great economies of scale if provided jointly.

Second, IMC would work where individual municipalities are left in control of primary functions within their jurisdiction. From a stakeholders' workshop which brought the three councils together as well as interviews held in individual councils, it was evident that councils would like to retain certain primary functions like policy making and service delivery in waste collection. This way, they feel assured that they are still in control or have autonomy over waste management within their jurisdiction. This means that cooperation would not be a welcome idea under waste collection despite the economies of scale. The stakeholders were however, in agreement that waste disposal and waste treatment would serve well for cooperation. Thus, in tandem with the multi-level governance perspective, some services can be handled in-house within a municipality while other services can be handled in cooperation. With the supporting legislation (The Local Government Act Cap 265) and possible models for cooperation (joint boards, contracting and under existing regional organisations) in Kenya, then certainly there is potential for IMC in solid waste management infrastructure provision which can be exploited to improve the SWM sector as a whole.

7.3.2 Cross country inter-municipal cooperation

Research question 4 addresses the role of regional organisations and networks in enhancing cross country SWM infrastructure provision amongst municipalities in the Lake Victoria Basin. To respond to this, the study looked at enabling and constraining factors for cooperation at the regional, national and municipal level, narrowed down to assessing the roles of two regional organisations that it categorises as statutory (formal) and voluntary (informal). As in section 7.3.1, interviews and two stakeholder workshops were key sources of data.

The study concludes that cross country municipal cooperation in infrastructure provision in the lake basin is not yet well developed. One would expect that with the shared resource, that is the lake basin, and the regional collaboration through the EAC as well as the flexibility of SWM, cooperation would be easily attained. From the interviews, observations and document analysis however, there is no evidence for cross country cooperation in SWM infrastructure. Though under a common framework efforts from these regional organisations are effected in municipalities individually in the different countries. The study links this outcome to the constraining factors at the national and municipal level. These constraining factors include the different levels of decentralisation in the three countries which in turn speak for differences in administrative, fiscal and political decentralisation to the municipal levels. The councils in the different countries have different autonomy levels over budget making, over staff recruitment while also the percentages of grants/funds that flow from the central government to the municipal councils are different. Cooperation in non-material aspects like information sharing and exchange visits is however, evident.

The formal and informal organisations can play different roles in enhancing cross country cooperation in SWM infrastructure provision. The formal organisation (LVBC) has so far played a supporting role to individual municipalities. Because of its strategic position with the three national governments and development partners, this organisation is seen to be able to play a key role in sourcing and mobilising resources for infrastructure provision. It also has the legal stability necessary for joint activities. The informal organisation (LVRLAC) has so far played a supplementary role for individual municipalities. Given its flexibility which allows local self-governing, it is seen to be able to play a key role in bringing the local authorities together on a common agenda on SWM particularly in the non-material aspects. The organisation mainly realises this by promoting information and knowledge sharing through joint stakeholders workshops or similar forums.

7.4 Concluding remarks

It is clear by now that SWM practices are governed differently in different countries and in turn in municipalities. The arrangements for service provision revealed by the research are outside the conventional or traditionally -recognised institutional arrangements and evolved into locally specific solutions. Even where the arrangement was formal, this was not implemented to its full extent, for instance the research did not come across a system that can be defined as fully franchised or a properly defined service/management contract. Instead the arrangements were structured by what seemed to work best locally. Resource mobilisation and allocation, management aspects and service provision are organised differently in different municipalities. Taking this reality into account therefore, means that modernising the institutional arrangement as affirmed by UN-Habitat (2010), will involve identifying, capitalising on, nurturing and improving on the local arrangements that are already working well in a particular locality. Modernised mixtures in urban solid waste management may be flexible, but they definitely need to build on existing practices in a particular context.

There is a possibility of a complete shift from municipal provision to other arrangements in the coming years. Nevertheless municipal authorities today are indeed at the centre of solid waste management in the East African urban authorities as presented in the conceptual models and they perceive of themselves that way. Although the overall levels of their performance are low, and legitimacy in the eyes of the citizens is not well established, the municipalities are reluctant to let go SWM as one of their key power resources. They recognise the need for improvement in many respects (following the solid waste chain) and not the least because they fear the damage done to legitimacy when there is a case of structural underperforming in SWM. While recognising the need for improving in a locally adopted way (modernised mixtures), some options for improving are more often tried/explored and recognised as feasible options than others. In general, formal relations are given priority over informal relations, and political public solutions are privileged over other (civil society/market based) solutions and strategies. As a result, the ideology of developmental state is reinforced while the notion of the network state is given less attention. The two cases of innovation (intermunicipal cooperation and international cooperation) made clear the constraining aspects of these new structures are given more emphasis by the actors involved than the new enabling aspects.

Against this overall background, the three criteria for SWM as provided by the MM-model are not (yet) met, and whether this will be improved in the near future remains to be seen.

References

ADB 2002. *Report Prepared for Sustainable development and Poverty Reduction Unit :Study On Solid Waste Management Options For Africa.* Côte d'Ivoire.

Adei, S. 2009. *State and Non-State Actors' Partnership and Collaboration: The Implications for Capacity Building in Public Service.* A paper presented at Africa Public Service Conference of Ministers of Public Service, Dar-es-Salaam, Tanzania, 17-18th June, 2009.

Adepoju, G.O. (ed.) 1999. *Managing the Monster: Urban waste and Governance In Africa.* Ottawa, Canada: IDRC.

Ahmad, E., Brosio, G. and Tanzi, V. 2008. *Local Service Provision in Selected OECD Countries: Do Decentralized Operation Work Better? IMF Working Paper- WP/08/67; pp. 1-35.* Washington, DC, USA: IMF.

Ahmed, J., Devarajan, S., Khemani, S. and Shah, S. 2005. *Decentralization and Service Delivery. World Bank Policy Research Working Paper No. 3603.* Washington, DC, USA: World Bank.

Ahmed, S.A. and Ali, S.M. 2006. People as Partners: Facilitating people's participation in public-private partnerships for solid waste management. *Habitat International,* 30 (4), 781-796.

Alderson, D. 2009. *Capacity Development for Quality Service Delivery at the Local Level in The Western Balkans. Municipal Status Reports on decentralized public service delivery for the municipalities participating in the implementation of the project Western Balkans on the Path to EU Integration: Strengthening Decentralized Service Delivery.* Croatia Country Office, Croatia: UNDP.

Anonymous, 2004/2005. Mwanza City Profile, 2004/2005. Available at: www.mwanza.de/index.

Anschutz, J. and Van de Klundert, A. 2000. *The Sustainability Of Alliances Between Stakeholders In Waste Management-using the concept of integrated Sustainable Waste Management - Working paper for UWEP/CWG.* The Netherlands: WASTE.

Ansell, C. 2000. The networked polity: Regional development in Western Europe. *Governance,* 13 (2), 279-291.

Arts, B., Leroy, P. and Tatenhove, J. 2006. Political modernization and Policy arrangement: A framework for understanding Environmental Policy Change. *Public Organization Review,* 6 (2), 93-106.

Atack, I. 1999. Four Criteria of Development NGO Legitimacy. *World Development,* 27(5), 855-864.

Awortwi, N. 2003. *Getting the fundamentals wrong: Governance of multiple modalities of basic services delivery in three Ghanaian Cities.* PhD Thesis. Rotterdam, the Netherlands: Institute of Social Studies.

Awortwi, N. 2003. *From Crisis in Urban Infrastructure Provision to Crisis in Governance: the management of multiple modalities in delivering public services in Ghanaian cities.* Paper presented at N-Aerus Annual Seminar in Paris, France - Beyond The Neo-Liberal Consensus on Urban Development: Other Voices from Europe and the South.

Azfar, O., Kahkonen, S., Lanyi, A., Meagher, P and Rutherford, D. 1999. *Decentralisation, Governance and Public Services: The Impact of Institutional Arrangement-A review of the Literature.* University of Maryland, MA, USA: IRIS centre.

Bach, S. 2000. *Decentralization and Privation in Municipal Services: The case of Health Services. Working paper No. 164.* Geneva, Switzerland: ILO.

Bartone, C.R. 2000. *Strategies for Improving Municipal Solid Waste Management: Lessons from World Bank Lending.* Paper presented at the International Workshop for Planning of Sustainable and Integrated Solid Waste Management, Manila, Philippines, 18-22 September 2000.

Baud, I.S.A., Post, J. and Furedy, C. 2004. *Solid waste management and recycling: actors, partnerships and policies in Hyderabad, India and Nairobi, Kenya.* The Netherlands: Kluwer Academic Publishers.

Bennett, E., Grohmann, P. and Gentry, B. 1999. Public-Private Partnerships for the Urban Environment: Options and Issues. *PPUE WP Series* Vol. 1. New York, NY, USA: UNDP Public-Private Partnerships for the Urban Environment and Yale University.

Boex, J. and Martinez-Vazquez, J. 2006. *Local Government Finance Reform in Developing Countries: The Case of Tanzania*. London: Palgrave Macmillan.

Braithwaite, V. and Levi, M. 1998. *Trust and Governance*. New York: Russell Sage Foundation

Bradford, N. 2008. *Multi-level Governance: Challenge & Opportunities for the Toronto City Region*. Canadian Policy Research Networks Publications.

Bratton, M. 1989. The politics of government-NGO relations in Africa. *World Development*, 17 (4), 569-587 In: I. Atack, 1999. Four Criteria of Development NGO Legitimacy. *World Development,* 27(5), 855-864.

Bulkeley, H., Watson, M., Hudson, R. and Weaver, P. 2005. Governing municipal waste: Towards a new analytical framework. *Journal of Environmental Policy & Planning*, 7 (1), 1- 23.

Caldwell, B.J., 2009 Centralization and Decentralization in Education: A New Dimension to Policy. *Globalization, Comparative Education and Policy Research*, 8 (1), 53-66.

Callaghy, T. 1993. Political Passions and Economic Interests; Economic Reform and Political Structure in Africa. In: Mkandawire, T. 1998. *Thinking About Developmental States in Africa*. A paper Presented at the UNU-AERC Workshop on Institution as and Development in Africa. Tokyo-Japan: UNU Headquarters

CAS Consultants for Ministry of local Government, Republic of Kenya, 2005. *Report on Feasibility Study on Solid Waste Management in Kisumu City Council, Kericho, Kisii, Nyamira and Homabay Municipal Councils*. Kenya: Nairobi

Central Bureau of Statistics 2003. *1999 Poverty Rates for rural Locations and Urban Sub-locations*. Nairobi: Government Printers.

Central Bureau of Statistics 2005. *1999 Poverty Rates for Constituencies*. Nairobi: Government Printers.

Clayton, A., Oakley, P. and Taylor, J. 2000. *Programme paper No. 2 on Civil society Organizations and Service Provision. Civil Society and social Movements*. Geneva, Switzerland: UNRISD.

CLGF and ComHabitat, 2005. *Report on Municipal Finance: Innovative Resourcing for Municipal Infrastructure and Service provision*. Coventry: Com Habitat.

Cointreau-Levine, S. 2005. *Notes on Solid Waste management Conceptual Issues on Cost recovery, Financial Incentives and Intergovernmental Transfers*. Available at: http://worldbank.org/solidwaste.

Cointreau-Levine, S. and Coad, A. 2002. *Guidance pack: Private Sector Participation in Municipal Solid Waste Management*. Switzerland: SKAT.

Conyers, D. 2007. Decentralization and Service Delivery: Lessons from Sub-Saharan Africa. *IDS Bulletin*, 38 (1), 18-32.

Conzelmann, T. 2008: A New Mode of Governing? Multi-level Governance between Cooperation and Conflict. In: Conzelmann, T. and Smith, R. (eds.), *Multi-level Governance in the European Union. Taking Stock and Looking Ahead*. Baden-Baden, Germany: Nomos, pp. 11-30.

Cotton, A.P., Sohail, M. and Tayler, W.K. 1998. *Community Initiatives in Urban Infrastructure*. Loughborough, UK: Loughborough University, Water Engineering and Development Centre.

Crook, R.C. and Manor, J. 1998. *Democracy and Decentralization in South Asia and West Africa*. Cambridge, UK: Cambridge University Press.

Crook, R.C. 2003. Decentralization and Poverty Reduction in Africa: The Politics of Central-Local Relations, *Public Administration and Development,* 23 (1),77-88.

Davies, A.R. 2003. Waste wars – public attitudes and the politics of place in waste management strategies. *Irish Geography*, 36(1), 77-92.

Davies, J. 2002. The governance of urban regeneration: a critique of the 'governing without government' thesis, *Public Administration*, 8(2), 301-322.

Davoudi, S. and Evans, N. 2005. The Challenge of Governance in Regional Waste Planning. *Environmental and Planning C: Government and Policy*, 23 (4), 493-517.

Da Zhu 2008. *Improving Municipal Solid Waste Management In India: A source book for Policy Makers and Practitioners.* Washington, DC, USA: The World Bank.

Devas, N. and Grant, U. 2003. Local government decision-making – citizen participation and local accountability: some evidence from Kenya and Uganda. *Public Administration and Development*, 23 (4), 307-316.

De Vaus, D. 2001. *Research Design in Social Research.* (1st Ed.). London: Sage Publication Ltd.

Develtere, P., Hertogen, E., and Wanyama, F. 2005. *The Emergence of Multi-level Governance in Kenya. Working Paper No. 7.* Leuven, Belgium: LIRGIAD K.U. Leuven. Avalaible at: www.lirgiad.be.

De Vries M.S. 2000. The Rise and fall of Decentralization: A Comparative Analysis of Arguments and Practices in European Countries. *European Journal of Political Research,* 38 (2),193-224.

DFID, 2005. Why we need to work more effectively in fragile states - DFID policy paper. London, UK: Department for International Development.

Dollery, B. 2005. *Alternative Approaches to Structural Reform in Regional and Rural Australian Local Government.* Paper presented to the North Queensland IPWEAQ Branch Conference in Bowen, Australia.

Douglas, M. 1966. *Purity and Danger.* London, UK: Routlegde & Kegan Paul.

Dryzek, J.S. 1997. The Politics of the Earth: Environmental Discourses. Reviewed by Seth Tuler in 1998. *Human Ecology Review*, 5 (1), 65-66.

EPA 1994. *Handbook on Joining Forces on Solid Waste Management: Regionalization is working in Rural and Small Communities; EPA530-K-93-001.*Washington, DC, USA: US Environmental Protection Agency.

EPA 2002. *Waste Transfer Stations: A Manual for Decision Making.* Washington, DC, USA: US Environmental Protection Agency.

Erie, S. and Mackenzie, S. 2007. *Regional Governance Reconsidered: Southern California Infrastructure.* Paper prepared for the annual meetings of the American Political Science Association. Chicago, IL, USA.

Farlam, P. 2005. *Assessing Public-Private Partnerships in Africa; Nepad Policy Focus Report No. 2.* South Africa: The South African Institute Of International Affairs.

Feiock, R.C. 2007. Rational Choice and Regional Governance. *Journal of Urban Affairs*, 29 (1), 47-63.

Fjeldstad, O. J. 2001. Taxation, Coercion and Donors: Local Government Tax Enforcement in Tanzania. *Journal of Modern African Studies,* 39 (2), 289-306.

Fjeldstad O.H. 2004. What has Trust Got To Do with It? Non-Payment of Service Charges in Local Authorities in South Africa. *Journal of Modern African Studies*, 42 (4), 539-562.

Gerber E.R. and Gibson C.C. 2005. *Cooperative Municipal Service Provision: A Political-Economy Framework for Understanding Intergovernmental Cooperation.* Political Science Working Group on Interlocal Services Cooperation. Paper presented at the Creating Collaborative Communities Conference. Detroit, MI, USA: Wayne State University.

Germa, B. and Xavier, F. 2010. Partial Privatization in Local Service Delivery: An Empirical Analysis of the Choice of Mixed Firms. *Local Government Studies*, 36 (1), 129-149.

Gibbs, D. 2000. Ecological Modernization, Regional Economic development and regional development agencies. *Geoforum*, 31(1), 9-19.

Gille, Z. 2007. *From the Cult of Waste to the Trash Heap of History* - The Politics of Waste in Socialist and Post Socialist Hungary. Indiana, USA: Indiana University Press.

Godsäter, A. and Söderbaum, F. 2008. *The Role of Civil Society in Regional Governance: The Case of Eastern and Southern Africa*. Paper for GARNET Annual Conference, Bordeaux, 17-19 September, 2008; the panel on 'The role of civil society in regional governance' (JERP 5.2.7: The Role of Non-state Actors and Civil Society in the Global Regulatory Framework).

Government of Uganda, undated. *Local Government Act Cap 243*. Kampala: LDC Publishers.

Government of Uganda, undated. *National Environment Act Cap 153*. Kampala: LDC Publishers.

Grafakos, S. and Baud, I. 1999. *Draft Synthesis Report on Alliances in urban environmental management: a search for indicators and contributions to sustainability*. Amsterdam, the Netherlands: AGIDS/IJWEP.

Guy, S., Marvin, S. and Moss, T. 2001. *Urban infrastructures in transition: networks, buildings, plans*. London, UK: Earthscan Publications Ltd.

Hansen, K. 1997. The Municipality between Central State and Local Self Government: Towards A New Municipality. *Local Government Studies*, 23 (4), 44-69.

Hanumm, H. and Lillich, R. 1980. The Concept of Autonomy in International Law. *American journal of International Law*, 74 (4), 858-889.

Harlow, C. and Rawlings, R. 2006. Promoting Accountability in Multi-Level Governance: A network Approach. *European Governance Papers No. C-06-02*. Available at: http://www.connex-network.org/eurogov/pdf/egf-connex-c-06-02.pdf.

Hegger, D. 2007. *Greening Sanitary Systems: An End-User Perspective*. PhD Thesis, Wageningen, the Netherlands: Wageningen University.

Helgøy, I., Homme, A. and Gewirtz, S. 2007. Introduction to Special Issue Local Autonomy or State Control? Exploring the Effects of New Forms of Regulation in Education. *European Educational Research Journal*, 6(3), 198-202.

Helmsing, A.H.J. 2002. Decentralization, Enablement and Local Governance in Low-Income Countries. *Environment and Planning C: Government and Policy*, 20 (3), 317-340.

Hulst, R. and Van Monfort A. 2007. The Netherlands: Cooperation as the Only Viable strategy. In: Hulst, R. and Van Monfort A. (eds.) *Intermunicipal Cooperation in Europe*. Berlin, Germany: Springer Publication, pp. 139-168.

ILO 2006. Employment Creation in Municipal Service Delivery, Improving Living Conditions and Providing Jobs For the Poor. *A Report from the Youth Employment and Urban Development Knowledge Sharing* Workshop, 27-28 February 2006, Johannesburg, South Africa.

JICA 2010. *Preparatory Survey for Integrated Solid Waste Management in Nairobi City in the Republic of Kenya*. Final Report. CTI engineering Co. Ltd and NJS Consultants Co. Ltd.

Jinja Municipal Council 2007. *Brief Profile of the Municipality*. Jinja: Uganda: Jinja Municipal Council.

Joardar, S.D. 2000. Urban Residential Solid Waste In India: Issues Related to Institutional Arrangements. *Public Works Management and Policy*, 4 (4), 319-330.

Johnson, C. 1999. The Developmental State: Odyssey of A Concept. In: Woo-Cumings, M. (ed.) *The Developmental State*. Ithaka, USA: Cornell University Press, pp. 32-60.

Jones, C., Herstely, W.S and Borgatti, S.P.1997. A General Theory of Network Governance: Exchange Conditions and Social Mechanisms. *The Academy of Management Review*, 22 (4), 911-945.

Jordan, B. 2005. What can communities and Institutions do? *Harvard International Review*, 27 (1), 1-4.

Jütting, J., Kauffmann, C., Mc Donnell, I., Osterrieder, H., Pinaud, N. and Wegner, L. 2004. *Decentralization and Poverty In Developing Countries: Exploring The Impact; Working Paper No. 236*. OECD Development Centre.

Karanja, A. 2005. *Solid waste management in Nairobi: actors, institutional arrangements and contribution to sustainable development*. PhD Dissertation, Rotterdam, the Netherlands: Institute of Social Studies.

Kennedy, D. 2002. *Contracting in Transition Economy Municipal Projects; Working paper No. 75.* London, UK: European Bank for Reconstruction and Development.

Kimenyi, M.S.2007. *Institutional Infrastructure to Support 'Super Growth' in Kenya: Governance Thresholds, Reversion Rates and Economic Development. Working Paper 2007-32.*

Kisumu Municipal Council. *Kisumu Municipal Council By-laws of 1954.* Kisumu, Kenya: Kisumu Municipal Council.

Klugman, J. 1994. *Occasional Paper No. 13 on Decentralization: A survey of literature from a Human Dimension Perspective.* New York, NY, USA: UNDP.

Krieckhaus, J. 2002. Reconceptualising The Developmental State: Public Savings and Economic Growth. *World Development*, 30 (10), 1697-1712.

Kumar, L. 2004. *Shifting Relationships Between the State and the Non-Profit Sector- Role of Contracts Under The New Governance Paradigm, Working Papers Vol. IV.* Toronto 6th International Conference: International Society For Third Sector Research.

Kwon, S.W. 2007. *Regional Governance Institutions and Interlocal Cooperation for Service Delivery. Working Group on Interlocal Services Cooperation.* Paper 28. Detroit, MI, USA: Wayne State University.

Langergraber, G., Meinzinger, F., Lechner, M., de Brujne, G., Sugden, S., Niwagaba, C.B., Mashauri, D., Mutua, B.M., Teklemariam, A., Achiro, I., Kiarie, S.C., Laizer, J.T. and Ayele, W. 2008. *A new approach to sustainable sanitation in Eastern African cities.* Proceedings of the 6th IWA World Water Congress, 8-12 September 2008, Vienna, Austria.

Lawrence, D.M. 2007. *Article No. 10 for County and Municipal Government in North Carolina on Interlocal Cooperation, Regional Organizations, and City County Consolidations.* Chapel Hill, NC, USA: UNC-Chapel Hill School of Government.

Lewis, P. 1996. Economic Reform and Political Transition in Africa: The Quest for a Politics of Development. *World Politics*, 49 (1), 92-129.

Lundqvist, L. 1987. Implementation Steering: an actor-structure approach. Lund, Sweden: Studentlitteratur.

Malunga, C. 2006. *Civil Society Efforts in Improving Accountability In Africa.* Paper presented at the ICCO Civil Society and Accountability Workshop, Harare, Zimbabwe.

Manus, C. 2005. *Report for LVWATSAN- Solid Waste Management Systems for Kisii and Homabay. UN Habitat Contract No. 4798.* Nairobi, Kenya.

Marks, G. and Hooghe, L. 2003. Unraveling the Central State, but How? Types of Multi-Level Governance. *American Political Science Review*, 97 (2), 233-243.

Marks, G. and Hooghe, L. 2004. Contrasting Visions of Multi-level Governance in: Ian Bache and Matthew Flinders (eds.). *Multi-Level Governance: Interdisciplinary Perspective.* Oxford, UK: Oxford University Press, pp. 15-30.

Matovu, G. 2002. Policy Options For Good Governance And Local Economic Development In Eastern And Southern Africa. *Urban Forum Journal*, 13 (4), 121-133.

McDonald, D. and Pape J. 2002. *Cost Recovery And The Crisis of Service Delivery in South Africa.* Pretoria, South Africa: HSRC Press.

Meijer, C.P., Verloop, N. and Beijaard, D. 2002. Multi-Method Triangulation in Qualitative Study on Teachers' Practical Knowledge: An attempt to Increase Internal Validity. *Quality and Quantity*, 36 (2), 145-167.

Menocal R.A. and Fritz, V. 2006. *(Re)building Developmental States: From Theory to Practice. Working Paper 274.* London, UK: Overseas Development Institute.

Mitullah, W.V. 2004. Participatory Governance for Poverty Alleviation in Local Authorities in Kenya: Lessons and Challenges. *Regional Development Dialogue,* 25 (1), 88-105.

Mmari. D.M.S. 2005. *The United Republic of Tanzania. Decentralization for Service Delivery in Tanzania.* A paper presented at the conference on Building Capacity for the Education Sector in Africa. Oslo, Norway.

Mkandawire, T. 1998. *Thinking About Developmental States in Africa*. A paper Presented at the UNU-AERC Workshop on Institution as and Development in Africa. Tokyo, Japan: UNU Headquarters.

Mkandawire, T. 2007. Good Governance: The itinerary of An Idea. *Development in Practice*, 17 (4-5), 679-683.

Modell, S. 2005. Triangulation between case study and survey methods in management accounting research: An assessment of validity implication. *Management Accounting Research*, 16 (2), 231-254.

Mol, A.P.J and Sonnenfeld, D. 2000. Ecological Modernization Around the World: An Introduction. *Environmental Politics, 9 (* 1), 3-16.

Mol, A.P.J, 2001. *Globalization and Environmental Reform. The Ecological Modernization of the Global Economy.* The MIT Press, Cambridge (MA).

Moyo, S.S., Kinuthia-Njenga, C. and UNCHS, 1998. *Privatization of municipal services in East Africa: A governance approach to human settlements management.* Nairobi: UN-Habitat.

Muchiri, E., Mutua, B. and Mullegger, E. 2010. Private Sector Involvement in Operating a Sanitation system with Urine Diversion Dry Toilets in Nakuru. In: *Sustainable Sanitation Practice*, Issue 2, 21-25.

Mutebi, F.G. 2008. Politics and Local Government in Uganda In: Saito, F. (ed.) 2008. *Foundations for Local Governance: Decentralization in Comparative Perspective* (pp137-164). Heidelberg: Physica-Verlag.

Mwangi, S.W. 2003. Challenge of urban environmental governance, Participation and Partnerships in Nakuru municipality, Kenya. PhD Thesis, Amsterdam, the Naetherlands: Amsterdam University.

Mwanza City Council, 2000. *Refuse Collection and Disposal By-laws of 2000*. Mwanza, Tanzania: Mwanza City Council.

Mwanza City Council 2007. *Mwanza city Brief profile*. Mwanza, Tanzania: Mwanza City Council.

Mwanza City Council, 2008. *Mwanza City Tender Document 2008/2009*. Mwanza, Tanzania: Mwanza City Council.

Mullin, M. 2007. *Do Special Districts Act Alone? Exploring the Relationship Between Flexible Boundaries and Intergovernmental Cooperation*. Paper prepared for the 2007 Annual Meeting of the Midwest Political Science Association, Chicago, IL, USA, 12-15 April 2007.

Newman, J. 2001. *Modernizing Governance: New Labour, policy and society*. London, UK: Sage Publications Ltd.

Noor, K.B.M. 2008. Case Study: A Strategic Research Methodology. *American Journal of Applied Sciences,* 5 (11), 1602-1604.

Norris, D.F. 2001. Prospects for Regional Governance under the New Regionalism: Economic Imperatives versus Political Impediments. *Journal of Urban Affairs*, 23 (5), 557-572.

Nsubuga-Kyaga, J. and Olum, Y. 2009. Local Governance and Local Democracy in Uganda. *Commonwealth Journal of Local Governance*, 2, 26-43.

Obirih-Opake, N. 2003. *Solid waste collection in Accra: Assessing the impact of decentralization and privatization on urban environmental management*. PhD Thesis Amsterdam, the Netherlands: Amsterdam University.

Okot-Okumu, J. 2006. *Solid Waste Management in Uganda: Issues Challenges and opportunities*. Paper presented at PROVIDE workshop in Wageningen, the Netherlands.

Okot-Okumu, J. and Oosterveer, P.J.M. 2010. Providing Sanitation For the Urban Poor in Uganda In: Van Vliet, B.J.M., Spaargaren, G. and Oosterveer, P. (eds.) *Social Perspectives on the Sanitation Challenge*. Dordrecht, the Netherlands,: Springer, pp. 49-66.

Olowu, D. 2002. Governance, Institutional Reforms and Policy Processes in Africa. In: Olowu, D. and Sako, S. (eds.) *Better Governance and Public policy: Capacity Building for democratic Renewal in Africa*. West Hartford, CT, USA: Kumarian Press Inc.

Olukoshi, A. 2005. *Changing Patterns of Politics in Africa. En libro: Politics and Social Movements in an Hegemonic World: Lessons from Africa, Asia and Latin America. Boron, Atilio A.; Lechini, Gladys.* CLACSO, Consejo Latinoamericano de Ciencias Sociales, Ciudad Autónoma de Buenos Aires, Argentina. Junio, pp: 177-201. Available at: http://bibliotecavirtual.clacso.org.ar/ar/libros/sursur/politics/Olukoshi.rtf.

Onyach-Olaa, M. 2003. The challenges of Implementing Decentralization: Recent Experience in Uganda. *Public Administration and development*, 23(1), 105-113.

Oosterveer, P. 2009. Urban environmental services and the state in East Africa; between neo-developmental and network governance approaches. *Geoforum*, 40 (6), 1061-1068.

Oosterveer, P. and Van Vliet, B.J.M. 2010. Environmental Systems and Local Actors: Decentralizing Environmental Policy in Uganda. *Environmental Management*, 45 (2), 284-295.

Oosterveer, P. and Spaargaren, G. 2010. Meeting Social Challenges in Developing Sustainable Environmental Infrastructures In: Van Vliet, B.J.M., Spaargaren, G. and Oosterveer, P. (eds.) *Social Perspectives on the Sanitation Challenge*. Dordrecht, the Netherlands,: Springer, pp. 11-30.

Ostrom, E. 1998. A Behavioral Approach to the Rational Choice Theory of Collective Action. *American Political Science Review*, 92(1), 1-22.

Oyugi, W.O. 2000. Decentralization for Good Governance and Development: The Unending Debate. *Regional Development Dialogue*, 21(1), iii-xix.

Parry, E., Kelliher, C., Mills, T. and Tison, S. 2005. Comparing HRM in the Voluntary and Public Sectors. *Journal of Personnel Review*, 34 (5), 588-602.

Pierre, J. and Peters, G. 2000. *Governance, Politics and the State*. London, UK: Macmillan Press Limited.

Provan, K.G. and Milward, B.H. 2001. Do networks really Work? A Framework For Evaluating Public Sector Organizational Networks. *Public Administration Review*, 61 (4), 414-423.

Provan, K.G. and Kenis, P. 2007. Modes of Network Governance: Structure, Management and Effectiveness. *Journal of Public Administration, Research and Theory*, 18 (2), 229-252.

Prud'homme, R. 1995. The Dangers of Decentralization. *World Bank Research Observer*, 10 (2), 201-220.

Pyndt, H. and Steffensen, J. 2005. 'Effective States and Engaged Societies: Capacity Development for Growth, Service Delivery, Empowerment and Security in Africa. In: *World Bank Review of Selected Experiences with Donor Support to Decentralisation in East Africa*', Local Government Denmark and Nordic Consulting Group.

Rakodi, C. 2003. *GISDECO Proceedings*. Politics and Performance: The Implications of Emerging Governance arrangements for Urban managements approaches and Information system. The Netherlands: International Institute for Geo-Information Science and Earth Observation. Available at: http://www.gisdevelopment.net/proceedings/gisdeco/sessions/key_karole.htm.

Ranja, T. 2004. *Privatization in East Africa: Gaps and Omissions in the Techniques. Report under ESRF Study on Globalization and the East African Economies*. Tanzania: Economic and Social Research Foundation.

Republic of Kenya 1989. *Income Tax Act Cap 470 of the Law of Kenya (1989)*. Nairobi, Kenya: Government Printers.

Republic of Kenya 1998 - *Local Government Act Cap 265*. Revised Edition. Nairobi, Kenya: Government Printers.

Republic of Kenya 2000. *Environmental Management and Co-ordination Act of 1999*. Nairobi-Government Printers.

Republic of Kenya 2001. *Migori District Development Plan 2002-2008*. Nairobi, Kenya: Government Printers.

Republic of Kenya 2001. *Homabay District Development Plan 2002-2008*. Nairobi, Kenya: Government Printers.

Republic of Kenya 2006. *Environmental Management and Co-ordination (Waste Management) Regulations, 2006*. Nairobi, Kenya: Government Printers.

Republic of Kenya 2006. *Finance Act (2006)*. Nairobi, Kenya: Government Printers.

Republic of Kenya 2010. *The Constitution of Kenya 2010*. Nairobi, Kenya: Government Printers.

Robinson, M. 2007. Does Decentralization Improve Equity and Efficiency in Public Service Delivery Provision? *Institute Of Development Studies Bulletin*, 38 (1), 7-17.

Robinson, M. and White, G. 1997. *The Role of Civic Organizations in the Provision of Social Services: Towards Synergy.* World Institute for Development Economics Research, Research For Action 37. Helsinki, Finland: UNU/WIDER.

Rondinelli D.A and Cheema, G.S. 1983. Implementing Decentralization Policies: An Introduction. In: Cheema G.S. and Rondinelli, D.A. (ed.) *Decentralization and Development: Policy Implementation in Developing Countries.* Beverly Hills, CA, USA: Sage Publications.

Rondinelli D.A., Nelli, J.R., and Cheema, G.S. 1984. *Decentralization in Developing Countries: A Review of Recent Experience. World Bank Staff, Working paper No. 581.* Washington, DC, USA: World Bank.

Rotich, K.H.; Yonsheng, Z. and Jun, D. 2006. Municipal Solid Waste Management Challenges in developing Countries-Kenyan Case Study. *Journal of Waste Management*, 26 (1), 92-100.

Saito, F. 2000. *Can Decentralization Contribute to 'Good Governance?': Lessons from Ugandan Experience.* A paper presented for the Annual Conference of the Development Studies Association, at the School of Oriental and African Studies, the University of London, UK.

Scheinberg, A. and Mol, A.P.J, 2010. Multiple modernities: Transitional Bulgaria and the Ecological modernization of solid waste management. *Environment and Planning C: Government and Policy*, 28 (1), 18-36.

Schmidt, R. and Quest Technology Inc., 2002. *Financing Regional Infrastructure.* NAIOP Research Foundation.

Schubeler, P. 1996. *Conceptual Framework for Municipal Solid Waste Management in Low Income Countries.* Working paper No. 9 of UMP/SDC collaborative programme on Municipal Solid Waste management. Switzerland: SKAT.

Smoke, P. 2001. *Fiscal Decentralization in Developing Countries: A review of Current concepts and Practice.* Democracy, Governance and Human Rights Programme Paper No. 2. Geneva, Switzerland: UNRISD.

Spaargaren, G., Oosterveer, P., Van Buren, J. and Mol, A.P.J. 2006. *Position Paper on Mixed modernities: Towards viable urban environmental infrastructure development in East Africa.* Wageningen, the Netherlands: Wageningen University.

Stamp, P. 1986. Local Government In Kenya: Ideology and Political Practice 1895-1974. *African Studies Review*, 29 (4), 17-42.

Steffensen, J., Tidemand, P., Naitore, H., Ssewankambo, E. and Mwaipopo, E. 2004. *Final synthesis report: A Comparative Analysis Of Decentralization In Kenya, Tanzania And Uganda.* Danish Trust Fund in the World Bank.

Steytler, N. (ed.), 2005. *The Place and Role of Local Government in Federal Systems - Occasional papers.* Johannesburg, South Africa: Konrad Adenauer Stiftung.

Strasser, S. 1999. *Waste and Want - A Social History of Trash.* New York, NY, USA: Henry Holt and Company LLC.

Stubbs, P. 2005. Stretching Concepts Too Far? Multi-Level Governance, Policy Transfer and the Politics of Scale in South East Europe. *Southeast European Politics*, VI (2), 66-87.

Tukahirwa, J.T., Mol, A.P.J. and Oosterveer, P. 2010. Civil society participation in urban sanitation and solid waste management in Uganda. *Local Environment*, 15 (1), 1-14.

Thurmaier, K. and Wood, C. 2002. Interlocal Agreements as Overlapping Social Networks: Picket-Fence Regionalism in Metropolitan Kansas City. *Public Administration Review,* 62(5), 585-598.

UNDP 2000. *Human Development Report, 2000 - World Population Projections 1998-2015.* New York, NY, USA: United Nations Development Programme.

UNDP 2010. *Capacities for Local Service Delivery: The Local Link. Global Event Working Paper.* New York, NY, USA: United Nations Development Programme.

UN-Habitat 2002. *Local Democracy and Decentralization in East And Southern Africa: Experiences from Uganda, Kenya, Botswana, Tanzania and Ethiopia.* Nairobi, Kenya: UN-Habitat.

UN Habitat 2003. *Water and sanitation in the world's Cities: Local action for global goals*. London, UK: Earthscan.

UN-Habitat 2008. *Baseline Survey Report on Kisumu Integrated Solid Waste Management*. Nairobi, Kenya: UN-Habitat.

UN-Habitat 2008. *Promoting Biodiversity In and Around lake Victoria basin*. Nairobi, Kenya: UN-Habitat.

UN-Habitat 2010. *Solid Waste Management In the World's Cities: Water and sanitation in the World's Cities 2010*. London, UK: Earthscan Ltd.

United Republic of Tanzania 1982. *Local Government (Urban Authorities) Act of 1982*. Dar-es salaam, Tanzania.

United Republic of Tanzania 1990. *National Health Policy of 1990*. Dar-es salaam, Tanzania.

United Republic of Tanzania 1997. *National Environmental policy of 1997*. Dar-es salaam, Tanzania.

United Republic of Tanzania 2002. *Public Service Act No.8 of 2002*. Dar-es salaam, Tanzania.

United Republic of Tanzania, 2004. *Environmental Management Act Cap 191*. Dar-es salaam, Tanzania.

USAID, 2006. *Making Cities Work - Urban Assessment Methodologies And Implementation Toolkit: Managing Municipal Services Delivery*. Washington, DC, USA: PADCO. Available at: http://www.makingcitieswork.org/urbanThemes/city_governance/service_delivery.

Van de Klundert, A., Cornelissen, M. and Scheinberg, A. 2004. *Report of Evaluation Mission to Lake Victoria*. Unpublished Report, WASTE and DGIS-DML.

Van Dijk, M.P. and Oduro-Kwarteng, S. 2007. Urban Management and Solid Waste Issues in Africa. *A contribution to the ISWA World congress in September 2007*. Amsterdam, the Netherlands.

Van Dijk, M.P., 2008. Urban Management and Institutional Change: An Integrated Approach to Ecological Cities. *IHS Working Papers Series 16/2008*. Rotterdam, the Netherlands.

Van Vliet, B.J.M. 2002. *Greening the grid, the ecological modernization of network bound systems*. PhD thesis. Wageningen, the Netherlands: Wageningen University.

Vedder, A. 2003. Non-State Actors' Interference in the International Debate on Moral Issues – Legitimacy and Accountability. In: Vedder, A. (ed.). *The WTO and Concerns Regarding Animals and Nature*. Nijmegen, the Netherlands: Wolf Legal Publishers, pp. 173-182.

Work, R. 2002. *Overview of Decentralization Worldwide: A Stepping Stone to Improved Governance and Human Development*. 2nd International Conference on Decentralization Federalism: The Future of Decentralizing States? Manila, Philippines.

World Bank 2000. *Cities in Transition: World Bank Urban and local Strategy*. Washington, DC, USA: The World bank Infrastructure Group Urban Development.

World Bank 2001. *The Philippines Environment Monitor 2001*. Washington, DC, USA: The World Bank.

World Bank 2009. *Improving Municipal Management For Cities to Succeed - An Independent Evaluation Group Special Study*. Washington, DC, USA: World Bank.

Wunsch, S.J. 2001. Decentralisation, Local Governance and 'Recentralization' In Africa. *Public Administration and Development*, 21(4), 277-288.

Yin, K.R. 1994 (2nd Ed), 2003 (3rd Ed.) and 2009 (4th Ed.). *Case Study Research: Design and Methods*. Newbury Park, CA, USA: Sage Publications.

Zurbrügg, C. 2003. *Urban Solid Waste Management in Low income countries of Asia - How to Cope with The Garbage Crisis*. Paper Presented for: Scientific Committee on Problems of the Environment (SCOPE) Urban Solid Waste Management Review Session, Durban, South Africa.

Annex 1. Household (estate) questionnaire

1. General information:
   Gender (male/female) ....
   Age ....
   Occupation ....
   Family size ....

2. Household income
   ○ Less than 5000
   ○ 5000-10,000
   ○ 10,000-20,000
   ○ More than 20,000

3. Who collects waste from your household?
   ○ Municipal ....
   ○ Private Co. ....
   ○ CBO/Youth group ....
   ○ Other (specify) ....

4. How many times a week? ....

5. Do they provide waste collection container? Yes/No
   Which type? ....
   And do you pay for the container? Yes/No
   How much? ....

6. How would you rate their service provision?
   ○ Satisfactory ....
   ○ Very satisfactory ....
   ○ Not satisfactory ....
   ○ Other (specify) ....

7. Do you pay for waste collection? Yes/No

8. How much? ....
   ....
   How did you reach an agreement on the amount to be paid? ....
   ....
   ....

9. How would you rate the amounts you pay for waste collection?
   - ○ Low
   - ○ Moderate
   - ○ High
   - ○ Other (specify)

10. Which service do you pay for:
    - ○ First ....
    - ○ Second ....
    - ○ Third ....
    - ○ Fourth ....
    Amongst Water & Sewerage, Electricity, Security and Waste Management.

11. Any suggestion for improving waste collection?
    ....
    ....
    ....
    ....

Annex 2. Interview guide for municipal authorities offices

1. Which department of the council deals with solid waste?

2. How is it organised? (organisation structure)

3. How many staff members are involved in waste management? (personnel)
   What are their qualifications?

4. What equipment does the council have for waste management?

5. What is the council budget for waste management?
   - Revenue from waste?
   - Expenditure to waste management?
   - Source of money spent on waste management?

6. Which regulations govern solid waste management?
   Are there any by-laws on waste management?

7. Which areas does the council serve in terms of waste collection?

8. How often is waste collected?

9. On average, how much waste is collected per day?

10. Any recycling initiatives by the council?

11. Does the council work in cooperation with other stakeholders in waste collection?
    If yes, who are these stakeholders?

12. What kind of cooperation?

13. Any kind of monitoring of these stakeholders?

14. What challenges does the council meet in waste management?
    Do the different levels (LCs) present a challenge in waste management?

15. What is the way forward to make better waste management?

Annex 3. Household responses on prioritising payments according to services offered

1st service	2nd service	3rd service	4th service
Kisumu (n=200)			
51% would pay for water as a first service	47% would pay for electricity as second service	76% would pay for electricity as third service	76% would pay for waste as last service
Mwanza (n=200)			
88% would pay for water as a first service	82% would pay for electricity as a second service	38% would pay for security as third service	76% would pay for waste as a last service
Jinja (n=218)			
61% would pay for water as a first service	31% would pay for electricity as a second service	31% indicated security as third services	52% would pay for waste as last service

Annex 4. List of selected recommendations from households for improving SWM

Jinja

- There is need to increase the frequency of collection at household level.
- There is need to explore other options like recycling.
- Increase the number of skips and bins to enhance proper disposal of solid wastes.
- There is need for monitoring of waste disposal at household level by the authorities.
- There is need to increase the capacity: staff and equipment for effective solid waste management.

Mwanza

- Areas selected as collection points should be fenced.
- Effective enforcement of laws/by-laws on waste management.
- Emphasize commitment by CBOs and private corporations towards waste management services.
- Employ adequate numbers of sanitation workers.
- Enhance community participation and involvement.
- Every 100 meters should a collection point be made.
- Every house should have a bucket for waste collection.
- Government should provide an adequate number of waste collection vehicles to enhance the process of waste collection.
- Government should provide modernised waste collection vehicles for waste collection.
- Government should improve infrastructure for areas that cannot access the waste collection services.
- Government should provide public education on waste management to the community.
- Improve infrastructures, increase the number of sanitation workers provide enough waste collection equipment.
- Increase the amount levied for waste collection.
- Increase the number of CBOs and private corporations and employ an adequate number of sanitation workers.
- More education on waste management and the amount paid for waste collection is also low.
- Municipal should provide refuse collection in the streets.
- Municipal to educate the community on waste management, collection fees, regular meetings with community members.
- Nobody should be allowed to burn waste unless permitted by the council.
- Not to bury wastes.
- Provide waste collection to areas not accessing the service.
- Provision of a recycling system at every collection point to help in waste management.
- Provision of waste collection service to areas that do not access the service.
- Punish those that do not pay for the waste collection services.
- Reduce the waste collection fees so that everyone can afford the service.
- The CBOs and private companies should improve cleanliness during night.
- The city council should engage the service of competent CBOs and private companies.

Kisumu

- There is need for better organisation among the SWM groups.
- The landlords should provide the service.
- There is need to increase the frequency of collection.
- There is need for competition to help better the services.
- Municipal council should provide the CBOs with transport.
- There is need to provide bins instead of polythene bags because they are not durable.
- Privatisation of SWM services.
- The system for SWM should be well structured.
- An estate wide approach should be adopted so as to cover as many households as possible.
- The CBOs and other groups should be funded since they do not have adequate facilities.
- The government should provide bins, skippers and eco-bags for free.
- The frequency of collection should be increased.

Annex 5. Local authority transfer fund[1] over the years in Homabay, Kisii and Migori

Year	Homabay MC	Kisii MC	Migori MC
FY 1999/2000	2,972,316	4,128,733	2,400,713
FY 2000/2001	7,062,893	5,920,255	8,510,357
FY 2001/2002	10,755,999	9,949,273	13,124,776
FY 2002/2003	10,755,999	9,949,273	13,124,776
FY 2003/2004	12,952,911	11,801,700	12,133,543
FY 2004/2005	13,816,804	12,568,444	12,941,159
FY 2005/2006	17,272,378	15,653,423	16,171,622
FY 2006/2007	25,911,313	23,302,870	24,247,780
FY 2007/2008	28,502,994	25,603,104	26,670,627
FY 2008/2009	31,958,568	28,670,083	29,901,089
Total allocation from 1999/2000 to FY 2008/2009	161,962,175	147,529,158	159,226,442

Source: Records from Association of Local Government Authorities in Kenya, 2009.

[1] Funds in Kshs.

Annex 6. Research-interview schedule for regional organisations

1. What is the organisation involved in generally?

2. Does organisation play any role in (solid) waste management in the Lake basin?
    What specific roles?

3. Which municipalities/councils are covered under the solid waste management?
    Why those municipalities? (criteria)

4. Which legislation and strategy govern their operations if any?

5. What is the organisation's:
    - budget;
    - equipment; and
    - staffing/personnel dedicated solely to waste management?

6. Possibilities for a regional approach in solid waste management amongst small urban councils?
    - Opportunities from an organisation point of view, and
    - Possible challenges?

7. What role is the organisation likely to play in such an arrangement?

8. Possibilities of the organisation participating in a workshop in Homabay, Kenya to steer a regional approach in January 2009?

Annex 7. Potential and existing areas of cooperation under SWM

Key elements	Approaches

Technical aspects (material flow)

Technologies	• Awareness creation to LAs on various technologies for SWM (e.g. bio-digesters, ECOSAN) • Capacity building of LA staff to implement the technologies • Acquiring technologies through partnerships with private sector and development agencies
Waste generation	• Sorting: waste generators should be responsible for their waste • Review and enforce by-laws • Promote sorting and 3Rs
Collection and Transport	• Common transport for recyclable material • Solicit common investment in local area for recycling purposes • Share trucks • Common contractor to transport the wastes • Promote private-public partnerships
Disposal	• Collaboration and cooperation to develop common sanitary landfill • Allow for LAs to use constructed landfill at a cost
Waste to wealth	• Sorting • Recycling • Composting

Non-technical aspects (non-material flow)

Innovations	• Market for the products made from waste materials • Improvement of production methods (capacity building) • Facilitate access to micro-credit scheme to support entrepreneurship development (availing information, sensitisation on loaning schemes)
Exchange study visits	• Networking • Promote good practices
Regulation	• Develop common regional LV body on SWM to oversee regulation/policies and harmonise them and later feed into the EAC
Networking	• Documentation and posting to common SWM website • Strengthen LVRLAC secretariat in SWM • LVRLAC monitoring mechanism in SWM
Participatory planning	• Public dialogue – bottom top planning • Gender mainstreaming
Community participation	• Setting aside public clean-up days • Organising cleanliness competitions (e.g. household level, village level, etc.)

Public private partnership	• Promoting partnership with the private sector
	• Encourage companies to fulfil corporate social responsibility
	• Forwarding of proposals for funding to the private sector
	• Making market connections
Community sensitisation	• Focal point discussions
	• Extension services
	• Public debates
	• IEC materials, visual aids like video shows, drama
Voluntary services	• Provision of incentives (certificates of appreciation)
	• Public appeals

Source: Information compiled from Bukoba and Kisumu Stakeholders Workshops in 2008 and 2009.

Annex 8. Interviewed key informants from different institutions

	Name	Department/organisation	Designation
1.	George Wesonga	LVRLAC Entebbe	Project Officer
2.	Julius Coredo Onyango	LVRLAC Kisumu	Project Officer
3.	Leah Oyake Ombis	Nairobi City Council	Director of Environment
4.	Benard Obera	Kisumu City Council	Director of Environment
5.	Harun Gulla	Kisumu City Council	Deputy director of Environment
6.	Zacheaus Okok	Kisumu City Council	Officer, Environmental Department
7.	Eutichus Okeyo Ochieng	Ten Stars, Tuungane, Kisumu	CBO Member
8.	Joseph Ochieng	SANA Kisumu	Officer
9.	Mathew Abuto	Public Health Department Migori Municipal Council	Public Health Officer
10.	Gerald Kibathi	NEMA-Migori Municipal Council	Environmental Officer
11.	Joseph Ayieko	Jangoe Youth Development Group-Migori	CBO Leader
12.	Paul Onyango Okello	Homabay Municipal Council	Town Engineer
13.	Mary Kerubo	Kisii Municipal Council	Head Public Health Officer
14.	Alphonse Mwanda	Kisii Municipal Council	Public Health Officer
15.	Gibson Gidudu	Jinja Municipal Council	Principal Health Officer
16.	Kigumba Wilberforce	Jinja Municipal Council-	Health Inspector
17.	Bituti Grace	Jinja Municipal Council	Health Inspector
18.	Beatrice Arigo	Jinja Municipal Council-	Contractor
19.	Ernest Nyavihanga	Jinja Municipal Council	Environment Officer
20.	Vicky Kakaire	Jinja Municipal Council	Environment Officer
21.	Audiphace Sabbo	Mwanza City Council	Public Health Inspector
22.	Dr. Kimaro	Mwanza City Council	City Director of Health
23.	Fredrick Salukele	Mwanza	PhD student in Mwanza

Summary

Solid Waste Management (SWM) as a subject has been extensively studied but taking the municipal authority as the focal point and using the Ecological Modernisation Theory to study it, gives it a new spur. Further approaching the study on a comparative basis of urban centres/municipalities that share an environmental resource in this case the Lake Victoria Basin and all having a common vision under the East African Community, attracts more attention to the subject matter. Impacting on the same shared resource – the Lake – this research sort to establish whether institutional arrangements of the different countries and in turn municipalities could generate insight into improving the plight of the urban poor as concerns management of solid waste.

Municipalities as the custodians of waste once its put out for collection have been reported numerously as performing poorly. This comes at a time when there is continuing debate on changing authority and shifting roles from the state to non-state actors. Yet it is also clear that municipalities have and continue to play an important role as an intermediary between the state and the public. Drawing upon the ecological modernisation theory, the study argues that 'modern' institutions (including the state/municipality; market and community) are key in bringing improvements to environmental infrastructures. This argument is presented through a number of theoretical discussions including the debates of centralised vs. decentralised systems of infrastructure and service provision; developmental state vs. the network governance arrangement and the theory of multi-level governance. The Modernised Mixtures Approach (MMA) is used to adapt Ecological Modernisation Theory for the study of SWM and sanitation. Under MMA, SWM systems have been deliberately and reflexively reconstructed in response to the challenge of a changing social, economic, and environmental context in East Africa. The modernised mixtures approach provides the criteria needed to assess performance of SWM infrastructure and service provision. The specific criteria include *accessibility* (particularly of the poor to avoid exclusion of particular groups): to what extent are specific groups included and excluded from environmental infrastructure due to financial, physical or cultural reasons?; *flexibility and resilience* (in both technological and institutional respect): how does the system or unit fit into more embracing future systems and how does it behave in times of instability in various dimensions (climate, political, economic, institutional)?; and *ecological sustainability* of the infrastructures and practices involved: to what extent do the new systems or the new technological options that become part of existing systems, improve the environmental performance of the urban infrastructure? These criteria are operationalised to reflect the local situation.

Therefore, to adequately make conclusions on the institutional arrangements, this study looked at:
- the status of the (physical) environmental infrastructures and the level of service provision for SWM in three designated urban centres in EA;
- the existing policy arrangement for SWM in those centres ;
- opportunities for cooperation in SWM amongst smaller neighbouring municipalities in the Lake basin;
- the role of regional organisations and networks in enhancing cross border SWM infrastructure provision.

Two conceptual frameworks helped to place the municipal authority at the centre within a nation and also at the regional level linking it to the other actors who are at different levels(household, neighbourhood, state and regional) and linking all these actors to the waste chain. The criteria of accessibility, flexibility and sustainability are then used to make assessments.

This study is based on case study research design. Multiple methods of data collection were used. Interviews with resource persons played a big role and provided data on the status of infrastructure, the policy arrangement, the possibility of IMC and also the role of regional organisations and networks in SWM. Household questionnaires were administered to gather data particularly on the relations between households and service providers as part of assessing the status of infrastructure and service provision as well as the policy arrangement.

Observation was helpful in providing data on the status of infrastructure. A stakeholders workshop brought neighbouring councils together to assess opportunities of IMC and the role of regional organisations and networks in SWM. Documents served as support for the information derived from other sources.

The status of infrastructure and service provision exhibited different performance levels amongst the three main municipalities studied. Kisumu showed a centralised system of infrastructure provision and management. The centralised system involves use of large scale infrastructure. The centralised arrangement resulted in only a section of the population receiving services. The type and amount of resources available were found inadequate. Mwanza and Jinja had a mixture of both centralised and decentralised systems where infrastructure and services came from both the municipality and the non-state actors which assured service provision to a bigger section of the population. The study thus concluded that SWM that incorporates the technical and social dimension reduces the centralised and decentralised dichotomy and in turn improves SWM.

The policy arrangement in the three municipalities depicted different array of actors which speaks for the relationship between municipalities and non-state actors and in turn their performance in SWM. In Kisumu, the involvement of non-state actors is informal and though the municipal authority is aware and even supports these actors, their informality affects their legitimacy as legal SWM service providers and in turn the performance of SWM. Jinja has formally incorporated private firms in form of contracts depicting an arrangement somewhat similar to a market arrangement. The services of these private firms are however, concentrated in the CBD where waste is more, given that they are paid per skip emptied. Areas away from the CBD do not receive frequent services. Mwanza has formally incorporated a mix of CBOs and private firms in form of contracts. CBOs are the majority depicting an arrangement referred to in the study as community dominated. This arrangement has afforded an almost 100% collection rate. In all the three councils however, the municipal authorities are still the loci of authority in SWM. Here the study concluded that a mixed arrangement that incorporates formal and informal actors with a mixture of actors from the state, community and private firms depicting a network arrangement as opposed to a developmental state of affairs, improves SWM infrastructure and service provision

The study further concluded that there are opportunities for inter-municipal cooperation amongst small neighbouring municipalities in Kenya in SWM. Their geographical proximity, combined population numbers, limited resources and low tax base in individual municipalities, the institutional arrangement and supporting legal framework as well as the amount of combined waste, present justifiable opportunities for cooperation. This is particularly the case in waste transportation, treatment and disposal as these present great economies of scale. The municipalities indicated the need for IMC that would leave certain primary functions within individual municipalities as a way of assuring them that they are still in charge of SWM and in turn autonomy over their jurisdictions. This pointed towards a multi-level governance arrangement.

The role of regional organisations and networks came out as both supportive and supplementary from the formal and informal regional organisations respectively. Their roles are faced by both constraining (the different decentralisation levels amongst the three East African Countries) and enabling factors(common legislation at the regional level and local level legislation that supports cooperation). Cooperation that cuts across the three countries was therefore evident in information sharing, capacity building and exchange of exemplary practices. Efforts at infrastructure provision by the regional organisation were however, limited to individual municipalities. The study concluded that cross country cooperation in SWM infrastructure provision is not yet well developed but that both the formal and informal regional organisations can play different roles in enhancing this cooperation.

The final general conclusion of this study is that while exemplary practices can be shared, institutional arrangements are typical to a specific location and therefore modernising SWM calls on identifying, capitalising on, nurturing and improving on the local arrangements that are already working well in a particular locality. Network-states options would result in better performing systems for SWM. This is because legitimacy of municipality would be increased while financial resources would be channelled towards market parties (and community organisations) under the control or supervision of the municipal authorities. Intermunicipal cooperation and international cooperation are two innovations in SWM for the case of East Africa urban authorities but attention needs to be shifted to factors that will enable the cooperation as opposed to focusing on constraining factors alone.

About the author

Christine Majale Liyala was born on 11th June 1979. She did her primary education (1989-1992) in Nangina girls primary school in Busia District, Kenya and proceeded to do her secondary education (1993-1996) at the Kenya High School in Nairobi, Kenya. She joined Kenyatta University for her Bachelor of Environmental Studies (1998-2000) and Masters of Environmental Planning and Management (2001-2003). After her Masters degree, the author worked with Kenya organisation for Environmental Education as a research assistant and was later employed at Kenyatta University as a tutorial fellow, where she works to date. In November 2006, the author joined Wageningen University for her PhD studies.

Printed in the United States
by Baker & Taylor Publisher Services